Theory of Nonneutral Plasmas

FRONTIERS IN PHYSICS: A Lecture Note and Reprint Series

David Pines, Editor

Volumes of the Series published from 1961 to 1973 are not officially numbered. The parenthetical numbers shown are designed to aid librarians and bibliographers to check the completeness of their holdings.

(1)	N. Bloembergen	Nuclear Magnetic Relaxation: A Reprint Volume, 1961
(2)	G. F. Chew	*S*-Matrix Theory of Strong Interactions: A Lecture Note and Reprint Volume, 1961
(3)	R. P. Feynman	Quantum Electrodynamics: A Lecture Note and Reprint Volume, 1961 (3rd printing, with corrections, 1973)
(4)	R. P. Feynman	The Theory of Fundamental Processes: A Lecture Note Volume, 1961 (2nd printing, with corrections, 1962)
(5)	L. Van Hove, N. M. Hugenholtz, and L. P. Howland	Problems in Quantum Theory of Many-Particle Systems: A Lecture Note and Reprint Volume, 1961
(6)	D. Pines	The Many-Body Problem: A Lecture Note and Reprint Volume, 1961 (3rd printing, with corrections, 1973)
(7)	H. Frauenfelder	The Mössbauer Effect: A Review—with a Collection of Reprints, 1962
(8)	L. P. Kadanoff and G. Baym	Quantum Statistical Mechanics: Green's Function Methods in Equilibrium and Nonequilibrium Problems, 1962 (2nd printing, 1971)
(9)	G. E. Pake	Paramagnetic Resonance: An Introductory Monograph, 1962 [cf. (42)—2nd edition]
(10)	S. C. Frautschi	Regge Poles and *S*-Matrix Theory, 1963
(11)	R. Hofstadter	Electron Scattering and Nuclear and Nucleon Structure: A Collection of Reprints with an Introduction, 1963
(12)	A. M. Lane	Nuclear Theory: Pairing Force Correlations to Collective Motion, 1963
(13)	R. Omnès and M. Froissart	Mandelstam Theory and Regge Poles: An Introduction for Experimentalists, 1963
(14)	E. J. Squires	Complex Angular Momenta and Particle Physics: A Lecture Note and Reprint Volume, 1963
(15)	H. L. Frisch and J. L. Lebowitz	The Equilibrium Theory of Classical Fluids: A Lecture Note and Reprint Volume, 1964
(16)	M. Gell-Mann and Y. Ne'eman	The Eightfold Way: (A Review—with a Collection of Reprints), 1964

FRONTIERS IN PHYSICS: A Lecture Note and Reprint Series

David Pines, Editor (*continued*)

(17)	M. Jacob and G. F. Chew	Strong-Interaction Physics: A Lecture Note Volume, 1964
(18)	P. Nozières	Theory of Interacting Fermi Systems, 1964
(19)	J. R. Schrieffer	Theory of Superconductivity, 1964 (2nd printing, 1971)
(20)	P. W. Anderson	Concepts in Solids: Lectures on the Theory of Solids, 1965 (2nd printing, 1971)
(21)	N. Bloembergen	Nonlinear Optics: A Lecture Note and Reprint Volume, 1965
(22)	R. Brout	Phase Transitions, 1965
(23)	I. M. Khalatnikov	An Introduction to the Theory of Superfluidity, 1965
(24)	P. G. de Gennes	Superconductivity of Metals and Alloys, 1966
(25)	W. A. Harrison	Pseudopotentials in the Theory of Metals, 1966 (2nd printing, 1971)
(26)	V. Barger and D. Cline	Phenomenological Theories of High Energy Scattering: An Experimental Evaluation, 1967
(27)	P. Choquard	The Anharmonic Crystal, 1967
(28)	T. Loucks	Augmented Plane Wave Method: A Guide to Performing Electronic Structure Calculations—A Lecture Note and Reprint Volume, 1967
(29)	Y. Ne'eman	Algebraic Theory of Particle Physics: Hadron Dynamics in Terms of Unitary Spin Currents, 1967
(30)	S. L. Adler and R. F. Dashen	Current Algebras and Applications to Particle Physics, 1968
(31)	A. B. Migdal	Nuclear Theory: The Quasiparticle Method, 1968
(32)	J. J. J. Kokkedee	The Quark Model, 1969
(33)	A. B. Migdal and V. Krainov	Approximation Methods in Quantum Mechanics, 1969
(34)	R. Z. Sagdeev and A. A. Galeev	Nonlinear Plasma Theory, 1969
(35)	J. Schwinger	Quantum Kinematics and Dynamics, 1970

(*continued*)

FRONTIERS IN PHYSICS: A Lecture Note and Reprint Series

David Pines, Editor (*continued*)

(36)	R. P. Feynman	Statistical Mechanics: A Set of Lectures, 1972 (3rd printing, 1974)
(37)	R. P. Feynman	Photon-Hadron Interactions, 1972
(38)	E. R. Caianiello	Combinatorics and Renormalization in Quantum Field Theory, 1973
(39)	G. B. Field, H. Arp, and J. N. Bahcall	The Redshift Controversy, 1973
(40)	D. Horn and F. Zachariasen	Hadron Physics at Very High Energies, 1973
(41)	S. Ichimaru	Basic Principles of Plasma Physics: A Statistical Approach, 1973
(42)	G. E. Pake and T. L. Estle	The Physical Principles of Electron Paramagnetic Resonance, 2nd Edition, completely revised, enlarged, and reset, 1973 [cf. (9)—1st edition]

Volumes published from 1974 onward are being numbered as an integral part of the bibliography:

Number		
43	R. C. Davidson	Theory of Nonneutral Plasmas, 1974
44	S. Doniach and E. H. Sondheimer	Green's Functions for Solid State Physicists, 1974
45	P. H. Frampton	Dual Resonance Models, 1974

THEORY OF Nonneutral Plasmas

Ronald C. Davidson
Department of Physics and Astronomy, University of Maryland, College Park

1974
W. A. BENJAMIN, INC.
ADVANCED BOOK PROGRAM
Reading, Massachusetts

London · Amsterdam · Don Mills, Ontario · Sydney · Tokyo

CODEN: FRPHA

Library of Congress Cataloging in Publication Data

Davidson, Ronald C
Theory of nonneutral plasmas.

(Frontiers in physics)
Bibliography: p.
1. Nonneutral plasma. I. Title. II. Series.
QC718.5.N4D38 530.4′4 73-17125

ISBN 0-8053-2345-7
ISBN 0-8053-2346-5 (pbk.)

Manufactured in the United States of America

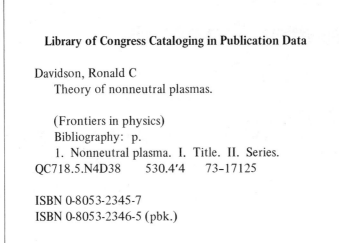

CONTENTS

EDITOR'S FOREWORD

The problem of communicating in a coherent fashion the recent developments in the most exciting and active fields of physics seems particularly pressing today. The enormous growth in the number of physicists has tended to make the familiar channels of communication considerably less effective. It has become increasingly difficult for experts in a given field to keep up with the current literature; the novice can only be confused. What is needed is both a consistent account of a field and the presentation of a definite "point of view" concerning it. Formal monographs cannot meet such a need in a rapidly developing field, and, perhaps more important, the review article seems to have fallen into disfavor. Indeed, it would seem that the people most actively engaged in developing a given field are the people least likely to write at length about it.

FRONTIERS IN PHYSICS has been conceived in an effort to improve the situation in several ways: first, to take advantage of the fact that the leading physicists today frequently give a series of lectures, a graduate seminar, or a graduate course in their special fields of interest. Such lectures serve to summarize the present status of a rapidly developing field and may well constitute the only coherent account available at the time. Often, notes on lectures exist (prepared by the lecturer himself, by graduate students, or by postdoctoral fellows) and have been distributed in mimeographed form on a limited basis. One of the principal purposes of the FRONTIERS IN PHYSICS Series is to make such notes available to a wider audience of physicists.

It should be emphasized that lecture notes are necessarily rough and informal, both in style and content, and those in the series will prove no exception. This is as it should be. The point of the series is to offer new, rapid, more informal, and it is hoped, more effective ways for physicists to teach one another. The point is lost if only elegant notes qualify.

The second way to improve communication in very active fields of physics is by the publication of collections of reprints of recent articles. Such collections are themselves useful to people working in the field. The value of the reprints would, however, seem much enhanced if the collection would be accompanied by an introduction of moderate length which would serve to tie the collection together and, necessarily, constitute a brief survey of the present status of the field. Again, it is appropriate that such an introduction be informal, in keeping with the active character of the field.

A third possibility for the series might be called an informal monograph, to connote the fact that it represents an intermediate step between lecture notes and formal monographs. It would offer the author an opportunity to present his views of a field that has developed to the point at which a summation might prove extraordinarily fruitful, but for which a formal monograph might not be feasible or desirable.

Fourth, there are the contemporary classics—papers or lectures which constitute a particularly valuable approach to the teaching and learning of physics today. Here one thinks of fields that lie at the heart of much of present-day research, but whose essentials are by now well understood, such as quantum electrodynamics or magnetic resonance. In such fields some of the best pedagogical material is not readily available, either because it consists of papers long out of print or lectures that have never been published.

The above words, written in August, 1961, seem equally applicable today. During the past decade, plasma physics has undergone a period of particularly rapid growth: today, it is an active, mature field of physics, containing a number of rapidly developing sub-fields, in many of which the current research effort is comparable to that for the field as a whole a decade ago. One such concerns the equilibrium stability properties of plasmas with large self fields; such magnetically confined nonneutral plasmas play an important role in collective-effect electron ring accelerators and in the generation and transport of intense high-current relativistic electron beams. In the present volume, Professor Davidson, himself an important contributor to the theory of nonneutral plasma, provides a lucid account of this important new sub-field of plasma physics, one which can be read with profit by both the experienced researcher and the graduate student beginning his research career.

DAVID PINES

Spring 1974

PREFACE

A *nonneutral* plasma is a many-body collection of charged particles in which there is *not* overall charge neutrality. Such systems can be characterized, depending on the charge density, by intense self electric fields. It has been known for some time that nonneutral plasmas exhibit *collective* properties that are qualitatively similar to those of neutral plasmas. For example, in klystrons and traveling-wave tubes, the collective oscillations necessary for microwave generation and amplification are excited even under conditions in which the electron beams in these devices are unneutralized.

The major recent interest in the equilibrium and stability properties of nonneutral plasmas originates from several diverse and rapidly developing research areas. These include (*a*) research on collective-effect accelerators (such as electron ring accelerators) that utilize the intense self fields of an electron cluster to trap and accelerate ions, (*b*) research on intense relativistic electron beams, with applications that include high-power microwave generation, ion acceleration in linear-beam geometries, and plasma heating via collective instabilities, and (*c*) studies of the stripping and confinement of heavy ions in nonneutral electron clouds in both toroidal and mirror magnetic field configurations. Although these research areas have different goals and objectives, they have in common the need to understand the equilibrium and stability properties of magnetically confined nonneutral plasmas that are characterized by intense self electric fields and (in high-current configurations) intense self magnetic fields.

This book is an introduction to the equilibrium and stability theory of magnetically confined nonneutral plasmas. Atomic processes and discrete particle interactions (i.e., binary collisions) are omitted from the analysis, and *collective* processes are assumed to dominate on the time and length scales of interest.

Extensive use is made of analytical techniques that are well established in the theory of neutral plasmas.

Two levels of theoretical description are available for a collisionless nonneutral plasma. These are (*a*) a *microscopic* or *kinetic* description based on the Vlasov-Maxwell equations, which includes finite-temperature effects in a natural manner, and (*b*) a *macroscopic* fluid description based on the moment-Maxwell equations. The basic equations and range of validity of the kinetic and macroscopic descriptions are summarized in Chapter 1. Chapter 2 deals with the macroscopic equilibrium and stability properties of *cold* nonneutral plasmas in uniform magnetic field geometries. The equilibrium and stability properties of magnetically confined nonneutral plasmas are examined within the framework of the Vlasov-Maxwell equations in Chapter 3. Configurations ranging from toroidal ring currents of relativistic electrons to intense relativistic electron beams in linear-beam geometries are analyzed.

RONALD C. DAVIDSON

College Park, Maryland

ACKNOWLEDGMENTS

This book on nonneutral plasma theory originated in a series of graduate lectures at the University of Maryland, and several people have contributed, directly or indirectly, to its content. I am especially grateful to Professor A. W. Trivelpiece and Dr. J. D. Lawson, who also lectured in the series, for numerous stimulating discussions that influenced the choice of theoretical topics included in this volume. I am also indebted to D. A. Hammer, N. A. Krall, P. C. Liewer, S. M. Mahajan, M. J. Schwartz, C. D. Striffler, and A. W. Trivelpiece for reading all or part of the manuscript and making valuable suggestions. In addition, I wish to thank Mrs. Mary Ann Ferg for her careful typing of the manuscript.

The work presented here has been supported in part by the National Science Foundation, in part by the Office of Naval Research, and in part by an Alfred P. Sloan Foundation Fellowship (1970–1972).

Finally, I thank my wife Jean for her gracious understanding and encouragement at every stage of this project.

Ronald C. Davidson

Theory of Nonneutral Plasmas

CHAPTER 1

INTRODUCTION

1.1 INTRODUCTION AND HISTORICAL BACKGROUND

This book deals with the equilibrium and stability theory of magnetically confined nonneutral plasmas. Atomic processes and discrete particle interactions (i.e., binary collisions) are omitted from the analysis, and *collective* processes are assumed to dominate on the time and length scales of interest.

A *nonneutral* plasma is a many-body collection of charged particles in which there is *not* overall charge neutrality. Such systems can be characterized, depending on the charge density, by intense self electric fields. It has been known for some time that nonneutral plasmas exhibit collective properties that are qualitatively similar to those of neutral plasmas. For example, microwave amplifying and generating devices, such as klystrons and traveling-wave tubes,[1] operate under high-vacuum conditions and depend on the existence and properties of collective oscillations[2] (space-charge waves) on drifting electron beams. For *continuous* operation it is reasonable to assume that the electron beams in these devices are electrically neutralized by the ions produced in ionizing collisions between the beam electrons and the low-density background gas. However, for *short-pulse* operation ($\sim$ 1 μsec, say) there is insufficient time for the ion density to build up to a significant level, and the electron beam is unneutralized. Nonetheless, in both cases the collective oscillations necessary for microwave generation and amplification are excited. Early experimental and theoretical studies of wave propagation along neutral and nonneutral magnetically focused electron beams[3–8] indeed indicated that overall charge neutrality is *not* a physical requirement for the existence of collective oscillations and shielding effects[9] in many-body charged particle systems.

In recent years there has been a considerable increase in interest in the equilibrium and stability properties of magnetically confined nonneutral plasmas. This interest is the result of several research programs, including the following:

1. The electron ring accelerator work at Berkeley,[10–14] Dubna,[15–18] Garching,[19] and Maryland.[20–25] The basic concept of a collective-effect accelerator[26, 27] utilizes the intense self fields of an electron cluster to trap and accelerate ions. This concept dates back more than two decades,[28–31] and experiments to form and transport such clusters were performed as early as 1952 by Alfvén and Wernholm.[28] The recent impetus to investigate intense relativistic electron rings as a suitable vehicle to trap and accelerate ions was provided by the extensive theoretical and experimental studies carried out by Veksler, Sarantsev, et al.[15, 16]
2. Experiments[32–43] to generate and transport intense high-current relativistic electron beams[44] in gaseous or plasma medium. Intense pulsed relativistic electron beams with power $> 10^{10}$ W have been used (or suggested for use) in various areas of research, such as microwave generation,[45, 46] thermonuclear fusion,[47–51] ion acceleration in linear-beam geometries,[52–55] and plasma heating [56, 57] via collective instabilities.[58–64]
3. The AVCO HIPAC studies of the acceleration and stripping of heavy ions in nonneutral electron clouds in toroidal magnetic fields.[65–70]
4. The Maryland studies of the fundamental equilibrium and stability properties of magnetically confined nonneutral plasmas in both mirror[71–75] and uniform[76–81] magnetic field geometries.
5. The Princeton studies of magnetoelectric confinement schemes for toroidal fusion plasmas,[82–84] and the stripping and confinement of heavy ions in nonneutral electron clouds in mirror magnetic fields.[85]

Although these programs have different goals and objectives, they have in common the need to understand the equilibrium and stability properties of magnetically confined nonneutral plasmas that are characterized by intense self electric fields and (in high-current configurations) intense self magnetic fields.

As stated earlier, this book deals with the equilibrium and stability theory of magnetically confined nonneutral plasmas. Atomic processes[86] and discrete particle interactions are omitted from the analysis, and *collective* processes are assumed to dominate on the time and length scales of interest. Two levels of theoretical description are available for a collisionless nonneutral plasma:

(*a*) a *microscopic* or *kinetic description* based on the Vlasov-Maxwell equations,[87–92] which includes finite-temperature effects in a natural manner; and
(*b*) a *macroscopic fluid description* based on the moment-Maxwell equations.[93]

The basic equations and range of validity of the kinetic and macroscopic descriptions are summarized in Section 1.3. As an introductory example to orient the reader and illustrate the dramatic effect that equilibrium self electric fields can have on particle trajectories, in Section 1.2 we examine the motion of an electron in a constant-density nonneutral plasma column aligned parallel to a uniform axial magnetic field $B_0\hat{\mathbf{e}}_z$.

Chapter 2 deals with the *macroscopic* equilibrium and stability properties of a *cold* nonneutral plasma column aligned parallel to a uniform axial magnetic field $B_0\hat{\mathbf{e}}_z$. The equilibrium state ($\partial/\partial t = 0$) is charcterized by a zero-order radial electric field. In the general case, both the mean azimuthal velocity and the mean axial velocity of the plasma components are allowed to be relativistic, and the corresponding axial and azimuthal self magnetic fields are included in the equilibrium analysis (Section 2.1). Various limiting equilibrium configurations are studied in Sections 2.2–2.4. These include equilibria in which the mean motions are nonrelativistic and the magnetic self fields are negligibly small (Section 2.2), equilibria in which the mean azimuthal motion is relativistic and the axial diamagnetic field is retained in the analysis (Section 2.3), and relativistic electron beam equilibria in which the mean axial motion is relativistic and the azimuthal self magnetic field is retained in the analysis (Section 2.4). In Section 2.5, a macroscopic equilibrium model of the Bennett pinch[44,94–96] is discussed, including the effects of finite beam temperature. Sections 2.6–2.10 deal with the macroscopic stability properties of nonneutral plasmas in the nonrelativistic regime. The analysis includes a study of stable electrostatic oscillations[8] analogous to those in a neutral plasma column[97] (Sections 2.7 and 2.8), electron-electron[79,80] and electron-ion[70] two-rotating-stream instabilities that result from the differential rotation of plasma components in the equilibrium radial electric field (Sections 2.8 and 2.9), and the diocotron instability[98–111] in hollow nonneutral electron beams (Section 2.10). Relativistic beam-plasma instabilities[59] are studied in Section 2.11.

Chapter 3 deals with the equilibrium and stability properties of magnetically confined nonneutral plasmas within the framework of the Vlasov-Maxwell equations. The general procedure for constructing self-consistent Vlasov equilibria for axisymmetric systems with equilibrium electric *and* magnetic self fields is discussed in Section 3.1. Several examples of specific equilibrium configurations are analyzed in Sections 3.2–3.5. These include nonrelativistic equilibria for a nonneutral plasma column[76–78] aligned parallel to a uniform axial magnetic field $B_0\hat{\mathbf{e}}_z$ (Section 3.2), relativistic E-layer equilibria[112–119] for Astron-like configurations[50,51,120,121] (Section 3.3), relativistic electron beam equilibria[122–127] in linear-beam geometries (Section 3.4), and relativistic electron ring equilibria[25,128–132] for a partially neutralized electron ring that is axially and radially confined in a mirror magnetic field (Section 3.5). An introduction to the Vlasov stability properties[76,77] of nonneutral plasmas is given in Sections 3.6 and 3.7.

1.2 CHARGED PARTICLE MOTION IN A NONNEUTRAL PLASMA COLUMN

As is the case with neutral plasmas, the macroscopic and Vlasov descriptions of nonneutral plasmas involve the use of averaged quantities such as the mean density or the distribution function. This means that in these models the details of the motion of individual charged particles are suppressed. There is a dramatic difference between the motion of charged particles in a neutral, field-free plasma and their motion in a nonneutral plasma, which has an equilibrium electric field. In order to examine this difference, the motion of a charged particle in a nonneutral plasma column is analyzed.

The simplest example of a nonneutral plasma is an infinitely long, unneutralized, constant-density cylindrical column of electrons immersed in a uniform axial magnetic field as shown in Fig. 1.2.1. No ions are present in the system. It is assumed that the electrons are not drifting parallel to the axis of symmetry, and that the diamagnetic field produced by the rotation of the plasma column about its axis of symmetry is negligible. It is also assumed that the thermal velocities of the electrons are negligible compared with their drift velocities in the equilibrium electric and magnetic fields, that is, the nonneutral plasma used in this example is "cold." The equilibrium state of this system can be investigated using the equations of motion for a charged particle in a steady radial electric field and a steady axial magnetic field.

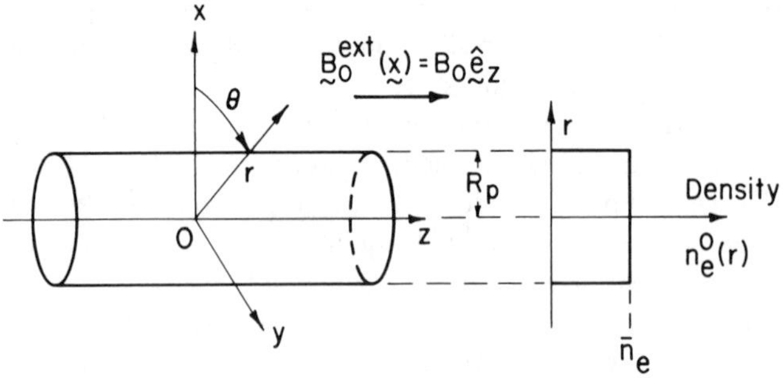

Fig. 1.2.1 Constant-density column of electrons immersed in a uniform axial magnetic field, $\mathbf{B}_0^{ext}(\mathbf{x}) = B_0\hat{\mathbf{e}}_z$.

For axicentered, circular, electron orbits, the radially outward centrifugal and electric forces on a given electron are balanced by the radially inward magnetic

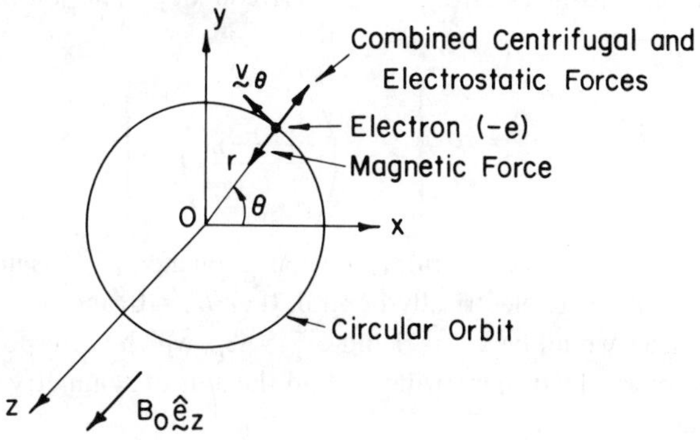

Fig. 1.2.2 Radial force balance for an axicentered, circular, electron orbit. The outward centrifugal and electric forces on the electron are balanced by the inward magnetic force [Eq. (1.2.1)].

force, as shown in Fig. 1.2.2. The radial force balance equation for an electron in a circular orbit is

$$-\frac{m_e v_{e\theta}^2}{r} = -eE_r^0 - ev_{e\theta}^0 B_0 . \qquad (1.2.1)$$

In Eq. (1.2.1), $v_{e\theta}^0(r)$ is the azimuthal velocity of the electron, $E_r^0(r)$ is the equilibrium radial electric field, and $-e$ and m_e are the electron charge and mass, respectively. The radial electric field is determined from Poisson's equation:

$$\frac{1}{r}\frac{\partial}{\partial r} rE_r^0 = -4\pi e n_e^0(r). \qquad (1.2.2)$$

For the constant-density profile shown in Fig. 1.2.1, Eq. (1.2.2) can be integrated to give $E_r^0 = -r4\pi e\bar{n}_e/2$ for $0<r<R_p$, which can be expressed in the equivalent form

$$E_r^0 = -\frac{m_e}{2e}\bar{\omega}_{pe}^2 r, \quad 0<r<R_p, \qquad (1.2.3)$$

where $\bar{\omega}_{pe}^2 \equiv 4\pi\bar{n}_e e^2/m_e$. Introducing the angular velocity $\omega_e = v_{e\theta}^0/r$, we can express Eq. (1.2.1) as

$$-\omega_e^2 = \frac{\bar{\omega}_{pe}^2}{2} - \omega_e\Omega_e, \qquad (1.2.4)$$

where $\Omega_e \equiv eB_0/m_e c$ is the electron cyclotron frequency. The solutions to Eq. (1.2.4) are

$$\omega_e = \omega_e^{\pm} \equiv \frac{\Omega_e}{2}\left[1 \pm \left(1 - \frac{2\overline{\omega}_{pe}^2}{\Omega_e^2}\right)^{1/2}\right]. \tag{1.2.5}$$

The fact that there are two equilibrium rotation frequencies[76,77] is not surprising. If the plasma column were electrically neutral, then $E_r^0 = 0$, and the equilibrium rotation frequencies would be $\omega_e^- = 0$ and $\omega_e^+ = \Omega_e$, which correspond to an electron at rest, or an electron gyrating around the axis of symmetry at the cyclotron frequency.

Since the plasma is assumed to be cold, the mean motion of the column is laminar. Therefore $\omega_e^{\pm}$ also represents the two possible mean rotation velocities of the column as a whole, that is, the motion of an individual electron is the same as that of a fluid element characterized by a mean equilibirum velocity $v_{e\theta}^0$. These two rotation frequencies ω_e^+ and ω_e^- are plotted versus $2\overline{\omega}_{pe}^2/\Omega_e^2$ in Fig. 1.2.3.

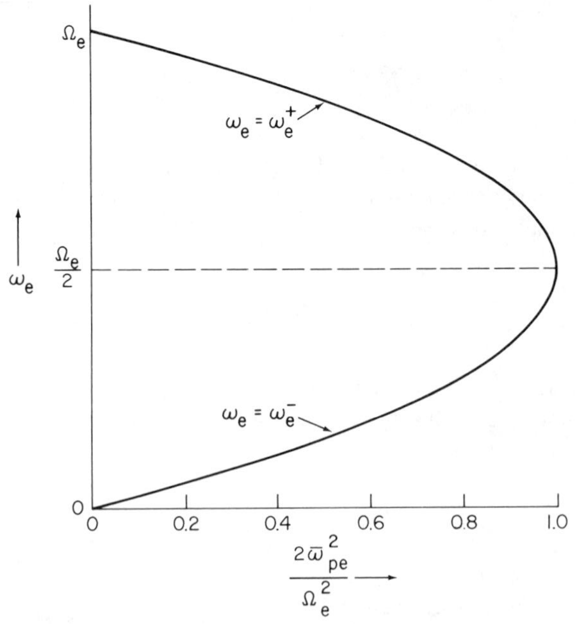

Fig. 1.2.3 Plot of the two allowed values of rotation frequency, ω_e^+ and ω_e^-, versus $2\overline{\omega}_{pe}^2/\Omega_e^2$ [Eq. (1.2.5)].

In the low density limit, $2\overline{\omega}_{pe}^2/\Omega_e^2 \ll 1$,

$$\omega_e^+ \simeq \Omega_e, \qquad \text{and} \quad \omega_e^- \simeq \frac{\overline{\omega}_{pe}^2}{2\Omega_e} = -\frac{cE_r^0}{rB_0}, \tag{1.2.6}$$

For low densities, the *fast* rotational mode ω_e^+ corresponds to all electrons gyrating around the axis of symmetry at the cyclotron frequency, whereas the *slow* rotational mode ω_e^- corresponds to an $\mathbf{E}^0 \times B_0\hat{\mathbf{e}}_z$ rotation of the column.

The high density limit, $2\overline{\omega}_{pe}^2/\Omega_e^2 = 1$, is known as Brillouin flow[133,134] for the case in which the beam is drifting along the magnetic field. It follows from Eq. (1.2.5) and Fig. 1.2.3 that

$$\omega_e^+ = \omega_e^- = \frac{\Omega_e}{2}, \quad \text{for} \quad \frac{2\overline{\omega}_{pe}^2}{\Omega_e^2} = 1. \tag{1.2.7}$$

The two rotation frequencies are equal and correspond to a rigid rotation of the column with angular velocity $\Omega_e/2$. From Eq. (1.2.5) it is seen that the condition

$$\frac{2\overline{\omega}_{pe}^2}{\Omega_e^2} \leqslant 1 \tag{1.2.8}$$

is required for $\omega_e^{\pm}$ to be real and thus to correspond to confined equilibrium solutions of the single-particle or cold-fluid equations. Equation (1.2.8) states that magnetic restoring forces (as measured by Ω_e^2) must overcome electrostatic repulsive forces (as measured by $\overline{\omega}_{pe}^2$) for radial confinement. If $2\overline{\omega}_{pe}^2/\Omega_e^2 > 1$, the beam expands radially, which is not an equilibrium situation.

The influence of a zero-order radial electric field E_r^0 on axicentered circular electron orbits is apparent for this simple example. (Keep in mind that $\omega_e = 0$ or Ω_e for a constant-density neutral plasma column.) Furthermore, it is shown in Chapters 2 and 3 that ω_e^+ and ω_e^- play an important role in the analysis of the stability properties of constant-density, nonneutral plasma columns.

The preceding analysis is readily extended to the case in which the electron orbit is not axicentered or circular. If a constant electron density profile is assumed as before, the equation of motion for an electron is

$$m_e\ddot{\mathbf{x}} = -e\mathbf{E}^0 - e\frac{\dot{\mathbf{x}} \times B_0\hat{\mathbf{e}}_z}{c}, \tag{1.2.9}$$

where $\mathbf{E}^0 = -(m_e/2e)\overline{\omega}_{pe}^2(x\hat{\mathbf{e}}_x + y\hat{\mathbf{e}}_y)$ in the column interior, and $\hat{\mathbf{e}}_x$ and $\hat{\mathbf{e}}_y$ are unit Cartesian vectors in a plane perpendicular to $B_0\hat{\mathbf{e}}_z$. The equations of motion for $x(t)$ and $y(t)$ are

$$\ddot{x} = \frac{\overline{\omega}_{pe}^2}{2} x - \Omega_e \dot{y}\,, \quad \ddot{y} = \frac{\overline{\omega}_{pe}^2}{2} y + \Omega_e \dot{x}\ . \tag{1.2.10}$$

It is useful to transform Eq. (1.2.10) to a frame of reference rotating with $\omega_e = \omega_e^+$ or $\omega_e = \omega_e^-$ (the two possible angular velocities of rotation of the column). Introducing

$$x' = x \cos \omega_e t + y \sin \omega_e t,$$

$$y' = y \cos \omega_e t - x \sin \omega_e t,$$

where x' and y' are the particle coordinates in the rotating frame gives, for Eq. (1.2.10),

$$\begin{aligned} \ddot{x}' &= -(\Omega_e - 2\omega_e)\dot{y}' + \left(\omega_e^2 - \Omega_e \omega_e + \frac{\overline{\omega}_{pe}^2}{2}\right) x', \\ \ddot{y}' &= (\Omega_e - 2\omega_e)\dot{x}' + \left(\omega_e^2 - \Omega_e \omega_e + \frac{\overline{\omega}_{pe}^2}{2}\right) y'. \end{aligned} \tag{1.2.11}$$

Since $\omega_e = \omega_e^{\pm}$ solves $\omega_e^2 - \Omega_e \omega_e + \overline{\omega}_{pe}^2/2 = 0$, the final terms in Eq. (1.2.11) vanish, giving

$$\begin{aligned} \ddot{x}' &= -(\Omega_e - 2\omega_e)\dot{y}', \\ \ddot{y}' &= (\Omega_e - 2\omega_e)\dot{x}'. \end{aligned} \tag{1.2.12}$$

Thus the particle motion as seen in the rotating frame consists of circular orbits with gyration frequency equal to the *vortex frequency*[8] ω_{ev}, defined by

$$\omega_{ev} = \Omega_e - 2\omega_e\,. \tag{1.2.13}$$

Since $\omega_e = \omega_e^+$ or $\omega_e = \omega_e^-$, it follows from Eqs. (1.2.5) and (1.2.13) that

$$|\omega_{ev}| = \omega_e^+ - \omega_e^- = \Omega_e \left(1 - \frac{2\overline{\omega}_{pe}^2}{\Omega_e^2}\right)^{1/2}. \tag{1.2.14}$$

For example, if an axicentered, circular electron orbit with radius r and angular velocity ω_e^- is perturbed, its motion in the rotating frame is circular with period $T = 2\pi/(\omega_e^+ - \omega_e^-)$, as shown in Fig. 1.2.4. However, the particle motion in the lab frame is trochoidal, as shown in Fig. 1.2.5.

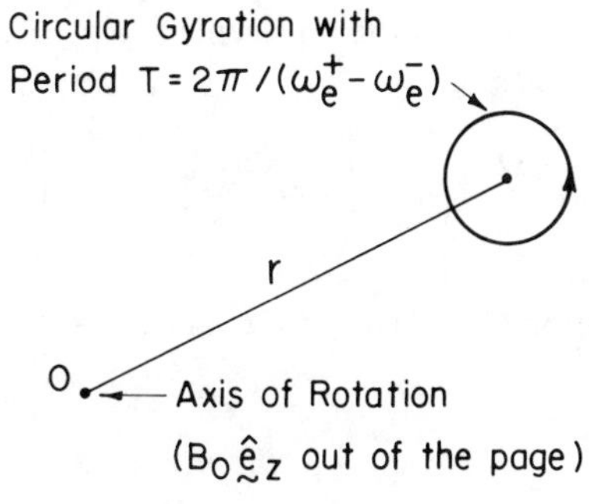

Fig. 1.2.4 The perturbed motion of an electron, viewed in a frame of reference rotating with angular velocity $\omega_e = \omega_e^-$. From Eqs. (1.2.12)-(1.2.14), the perturbed motion in the rotating frame is a circular gyration with period $T = 2\pi/(\omega_e^+ - \omega_e^-)$. If $\omega_e = \omega_e^+$, the sense of gyration is opposite to that shown in the figure.

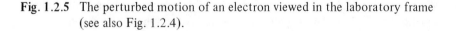

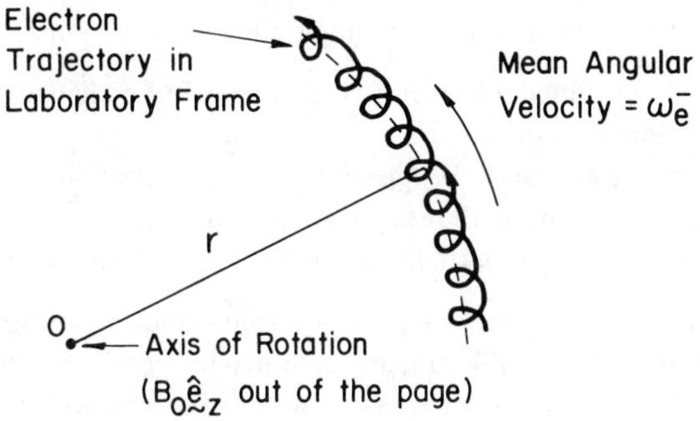

Fig. 1.2.5 The perturbed motion of an electron viewed in the laboratory frame (see also Fig. 1.2.4).

The preceding analysis is readily extended to include a fixed (infinitely massive), partially neutralizing, ion background with density

$$n_i^0 = fn_e^0, \tag{1.2.15}$$

where f = const. = fractional neutralization. If single ionization is assumed, the only modification in the analysis is the replacement $\bar{\omega}_{pe}^2 \to \bar{\omega}_{pe}^2 (1-f)$. For example, Eq. (1.2.5) becomes

$$\omega_e = \omega_e^{\pm} = \frac{\Omega_e}{2}\left[1 \pm \left(1 - \frac{2\bar{\omega}_{pe}^2}{\Omega_e^2}(1-f)\right)^{1/2}\right]. \qquad (1.2.16)$$

For complete neutralization ($f = 1$), $\omega_e^+ = \Omega_e$ and $\omega_e^- = 0$, as expected.

1.3 LEVELS OF THEORETICAL DESCRIPTION

1.3.1 General Discussion

The term *nonneutral plasma* is used to describe a system of charged particles in which there is *not* overall charge neutrality. Such systems are characterized by zero-order equilibrium electric fields, $\mathbf{E}^0(\mathbf{x})$, which are ordinarly absent in neutral plasmas. Chapters 2 and 3 include studies of the equilibrium and stability properties of the following nonneutral plasma systems:

1. Magnetically confined, electron-rich plasma column aligned parallel to a uniform external magnetic field.
2. Relativistic electron beam propagating through a partially neutralizing ion background, with and without a magnetic guide field.
3. Magnetically confined, partially neutralized, relativistic electron rings.

It is assumed that the nonneutral plasma systems under investigation are *collisionless*, that is, the equilibrium and stability properties of these systems are studied for time scales *short* compared with a binary collision time. As stated in Section 1.1, two levels of theoretical description of a collisionless plasma are at our disposal: a *macroscopic fluid description* based on moment-Maxwell equations,[93] and a *microscopic description* based on Vlasov-Maxwell equations.[87–92] Both levels of description are used in subsequent chapters to study the properties of nonneutral plasmas.

In a *macroscopic fluid description* we examine the time development of macroscopic properties of the plasma, such as

$n_\alpha(\mathbf{x},t)$ = number density of the αth plasma component,

$\mathbf{V}_\alpha(\mathbf{x},t)$ = mean velocity of the αth plasma component,

$\boldsymbol{P}_\alpha(\mathbf{x},t)$ = pressure tensor for the αth plasma component.

These quantities evolve self-consistently in terms of the electric and magnetic fields, $\mathbf{E}(\mathbf{x}, t)$ and $\mathbf{B}(\mathbf{x}, t)$, determined from Maxwell's equations. The advantage of such a description is its simplicity. If the plasma is *cold*, variations in the pressure can be ignored and thus the approximation

$$\frac{\partial}{\partial \mathbf{x}} \cdot \boldsymbol{P}_\alpha \simeq 0 \tag{1.3.1}$$

can be made. This approximation results in a closed description of the time development of $n_\alpha(\mathbf{x}, t)$, $\mathbf{V}_\alpha(\mathbf{x}, t)$, $\mathbf{E}(\mathbf{x}, t)$, and $\mathbf{B}(\mathbf{x}, t)$, based on the continuity equation, equation of motion for the fluid, and Maxwell's equations. Both equilibrium and stability properties can be discussed using such a model. Since the description is macroscopic, the stability properties of course depend on gross features of the equilibrium, such as equilibrium density and velocity profiles, $n_\alpha^0(\mathbf{x})$ and $\mathbf{V}_\alpha^0(\mathbf{x})$. This description of a nonneutral plasma is useful because of its simplicity. Finite geometry effects can be handled with relative ease in a macroscopic cold-plasma description. There are, however, two main objections to such a fluid approach. First, it is not straightforward to extend a cold-fluid model to include finite temperature effects, that is, it is not generally known what equations of state to use for the stress tensor $\boldsymbol{P}_\alpha(\mathbf{x}, t)$. Second, certain phenomena, such as Landau damping[135] and waves and instabilities associated with the detailed momentum-space distribution of plasma particles, cannot be investigated using the macroscopic fluid description of a plasma, either neutral or nonneutral.

To include the effects of finite temperature on the equilibrium and stability of nonneutral plasmas, it is necessary to study them within a *kinetic* (Vlasov-Maxwell) framework. In this case, the one-particle distribution function,† $f_\alpha(\mathbf{x}, \mathbf{p}, t)$, and average electric and magnetic fields, $\mathbf{E}(\mathbf{x}, t)$ and $\mathbf{B}(\mathbf{x}, t)$, evolve self-consistently according to the Vlasov-Maxwell equations. Self-consistent equilibria are readily constructed in such a framework. Also there is a broad class of plasma waves and instabilities that depend on the detailed $\mathbf{p}$-space structure of the equilibrium distribution $f_\alpha^0(\mathbf{x}, \mathbf{p})$. Such waves and instabilities cannot be analyzed by means of a macroscopic cold-fluid plasma model. Although a broad class of nonuniform equilibria can be constructed using the Vlasov-Maxwell equations, it should be pointed out that a stability analysis based on them is generally more complicated than one based on a macroscopic fluid description.

For future reference the basic equations used in Chapters 2 and 3 to describe the equilibrium and stability properties of collisionless nonneutral plasma are summarized below.

---

†Here $f_\alpha(\mathbf{x}, \mathbf{p}, t)\, d^3x\, d^3p$ is the probable number of particles of component α with position $\mathbf{x}$ and momentum $\mathbf{p}$ in the interval $d^3x\, d^3p$ at time t.

1.3.2 The Vlasov Description

The one-particle distribution function in configuration-momentum space, $f_\alpha(\mathbf{x}, \mathbf{p}, t)$, evolves according to the *relativistic* Vlasov equation,

$$\left[\frac{\partial}{\partial t} + \mathbf{v} \cdot \frac{\partial}{\partial \mathbf{x}} + e_\alpha \left(\mathbf{E} + \frac{\mathbf{v} \times \mathbf{B}}{c}\right) \cdot \frac{\partial}{\partial \mathbf{p}}\right] f_\alpha(\mathbf{x}, \mathbf{p}, t) = 0, \tag{1.3.2}$$

where $\mathbf{v}$ and $\mathbf{p}$ are related by

$$\mathbf{v} = \frac{\mathbf{p}/m_\alpha}{(1 + \mathbf{p}^2/m_\alpha^2 c^2)^{1/2}} . \tag{1.3.3}$$

The electric and magnetic fields, $\mathbf{E}(\mathbf{x}, t)$ and $\mathbf{B}(\mathbf{x}, t)$, in Eq. (1.3.2) are determined self-consistently from Maxwell's equations,

$$\nabla \times \mathbf{E} = -\frac{1}{c}\frac{\partial \mathbf{B}}{\partial t}, \tag{1.3.4}$$

$$\nabla \times \mathbf{B} = \frac{4\pi}{c} \sum_\alpha e_\alpha \int d^3p\, \mathbf{v} f_\alpha(\mathbf{x}, \mathbf{p}, t) + \frac{4\pi}{c}\mathbf{J}_{\text{ext}} + \frac{1}{c}\frac{\partial \mathbf{E}}{\partial t}, \tag{1.3.5}$$

$$\nabla \cdot \mathbf{E} = 4\pi \sum_\alpha e_\alpha \int d^3p\, f_\alpha(\mathbf{x}, \mathbf{p}, t) + 4\pi\rho_{\text{ext}} , \tag{1.3.6}$$

$$\nabla \cdot \mathbf{B} = 0. \tag{1.3.7}$$

Equations (1.3.6) and (1.3.5) allow the possibility of external charge and current sources, $\rho_{\text{ext}}(\mathbf{x}, t)$ and $\mathbf{J}_{\text{ext}}(\mathbf{x}, t)$. Equation (1.3.2) is Liouville's theorem for the incompressible motion of component α in the six-dimensional phase space $(\mathbf{x}, \mathbf{p})$. Note that Eq. (1.3.2) is nonlinear since $\mathbf{E}(\mathbf{x}, t)$ and $\mathbf{B}(\mathbf{x}, t)$ are determined self-consistently in terms of $f_\alpha(\mathbf{x}, \mathbf{p}, t)$ from Maxwell's equations.

An *equilibrium analysis* of Eq. (1.3.2) and Eqs. (1.3.4)-(1.3.7) proceeds by setting $\partial/\partial t = 0$ and looking for stationary solutions, $f_\alpha^0(\mathbf{x}, \mathbf{p})$, $\mathbf{E}^0(\mathbf{x})$, and $\mathbf{B}^0(\mathbf{x})$, that satisfy† the equations

$$\left[\mathbf{v} \cdot \frac{\partial}{\partial x} + e_\alpha \left(\mathbf{E}^0 + \frac{\mathbf{v} \times \mathbf{B}^0}{c}\right) \cdot \frac{\partial}{\partial \mathbf{p}}\right] f_\alpha^0(\mathbf{x}, \mathbf{p}) = 0, \tag{1.3.8}$$

$$\nabla \times \mathbf{E}^0 = 0, \tag{1.3.9}$$

† In writing Eqs. (1.3.10) and (1.3.11) it has been assumed that $\mathbf{J}_{\text{ext}}$ and ρ_{ext} are independent of t.

$$\nabla \times \mathbf{B}^0 = \frac{4\pi}{c} \sum_\alpha e_\alpha \int d^3p\, \mathbf{v} f_\alpha^0(\mathbf{x}, \mathbf{p}) + \frac{4\pi}{c} \mathbf{J}_{\text{ext}}(\mathbf{x}), \tag{1.3.10}$$

$$\nabla \cdot \mathbf{E}^0 = 4\pi \sum_\alpha e_\alpha \int d^3p\, f_\alpha^0(\mathbf{x}, \mathbf{p}) + 4\pi\rho_{\text{ext}}(\mathbf{x}), \tag{1.3.11}$$

$$\nabla \cdot \mathbf{B}^0 = 0. \tag{1.3.12}$$

An analysis of Eq. (1.3.8) reduces to a determination of the single-particle constants of the motion in the equilibrium fields $\mathbf{E}^0(\mathbf{x})$ and $\mathbf{B}^0(\mathbf{x})$.[137] For the applications of interest here, $\mathbf{E}^0(\mathbf{x})$ is produced by deviations from charge neutrality in equilibrium [i.e., $\rho_{\text{ext}}(\mathbf{x}) = 0$, but $\Sigma_\alpha e_\alpha \int d^3p\, f_\alpha^0(\mathbf{x}, \mathbf{p}) \neq 0$], and $\mathbf{B}^0(\mathbf{x})$ is produced by external current sources as well as any equilibrium plasma currents, for example, a relativisic electron beam passing through a fixed ion background. The term *equilibrium* as used here should not be confused with *thermal equilibrium*. For a given external field configuration there can, in general, be many *Vlasov equilibria*. These equilibria are stationary states that can exist on a time scale less than a binary collision time.[137] A specific equilibrium may be unstable if perturbations about the equilibrium grow in time or space.

A stability analysis based on Eqs. (1.3.2)-(1.3.7) proceeds in the following manner. The quantities $f_\alpha(\mathbf{x}, \mathbf{p}, t)$, $\mathbf{E}(\mathbf{x}, t)$, and $\mathbf{B}(\mathbf{x}, t)$ are expressed as the sum of their equilibrium values plus a time-dependent perturbation:

$$f_\alpha(\mathbf{x}, \mathbf{p}, t) = f_\alpha^0(\mathbf{x}, \mathbf{p}) + \delta f_\alpha(\mathbf{x}, \mathbf{p}, t),$$

$$\mathbf{E}(\mathbf{x}, t) = \mathbf{E}^0(\mathbf{x}) + \delta\mathbf{E}(\mathbf{x}, t),$$

$$\mathbf{B}(\mathbf{x}, t) = \mathbf{B}^0(\mathbf{x}) + \delta\mathbf{B}(\mathbf{x}, t). \tag{1.3.13}$$

The quantities $f_\alpha^0(\mathbf{x}, \mathbf{p})$, $\mathbf{E}^0(\mathbf{x})$, and $\mathbf{B}^0(\mathbf{x})$ satisfy Eqs. (1.3.8)-(1.3.12). The time development of the perturbations $\delta f_\alpha(\mathbf{x}, \mathbf{p}, t)$, $\delta\mathbf{E}(\mathbf{x}, t)$, and $\delta\mathbf{B}(\mathbf{x}, t)$ is studied using Eqs. (1.3.2)-(1.3.7). For *small-amplitude* perturbations, the Vlasov-Maxwell equations are *linearized* about the equilibrium $f_\alpha^0(\mathbf{x}, \mathbf{p})$, $\mathbf{E}^0(\mathbf{x})$, and $\mathbf{B}^0(\mathbf{x})$. This gives

$$\left[\frac{\partial}{\partial t} + \mathbf{v} \cdot \frac{\partial}{\partial \mathbf{x}} + e_\alpha \left(\mathbf{E}^0 + \frac{\mathbf{v} \times \mathbf{B}^0}{c}\right) \cdot \frac{\partial}{\partial \mathbf{p}}\right] \delta f_\alpha(\mathbf{x}, \mathbf{p}, t)$$
$$= -e_\alpha \left(\delta\mathbf{E} + \frac{\mathbf{v} \times \delta\mathbf{B}}{c}\right) \cdot \frac{\partial}{\partial \mathbf{p}} f_\alpha^0(\mathbf{x}, \mathbf{p}), \tag{1.3.14}$$

$$\nabla \times \delta\mathbf{E} = -\frac{1}{c}\frac{\partial}{\partial t}\, \delta\mathbf{B}, \tag{1.3.15}$$

$$\nabla \times \delta \mathbf{B} = \frac{4\pi}{c} \sum_\alpha e_\alpha \int d^3p\, \mathbf{v}\, \delta f_\alpha(\mathbf{x}, \mathbf{p}, t) + \frac{1}{c}\frac{\partial}{\partial t} \delta \mathbf{E}, \tag{1.3.16}$$

$$\nabla \cdot \delta \mathbf{E} = 4\pi \sum_\alpha e_\alpha \int d^3p\, \delta f_\alpha(\mathbf{x}, \mathbf{p}, t), \tag{1.3.17}$$

$$\nabla \cdot \delta \mathbf{B} = 0, \tag{1.3.18}$$

where $\mathbf{v}$ and $\mathbf{p}$ are related by Eq. (1.3.3). If the perturbations $\delta f_\alpha(\mathbf{x}, \mathbf{p}, t)$, $\delta \mathbf{E}(\mathbf{x}, t)$, and $\delta \mathbf{B}(\mathbf{x}, t)$ grow, the equilibrium distribution $f_\alpha^0(\mathbf{x}, \mathbf{p})$ is *unstable.* If the perturbations damp, the system returns to equilibrium and is *stable*. For spatially nonuniform equilibria with space charge, a stability analysis based on Eqs. (1.3.14)-(1.3.18) is generally mathematically formidable.

1.3.3 The Macroscopic Fluid Description

In this section, the essential features of a macroscopic plasma description[93] based on the moment-Maxwell equations are summarized. The αth-component particle density, $n_\alpha(\mathbf{x}, t)$, mean velocity, $\mathbf{V}_\alpha(\mathbf{x}, t)$, mean momentum, $\mathbf{P}_\alpha(\mathbf{x}, t)$, and pressure tensor, $\boldsymbol{P}_\alpha(\mathbf{x}, t)$, are defined as follows:

$$n_\alpha(\mathbf{x}, t) \equiv \int d^3p\, f_\alpha(\mathbf{x}, \mathbf{p}, t), \tag{1.3.19}$$

$$n_\alpha(\mathbf{x}, t)\mathbf{V}_\alpha(\mathbf{x}, t) \equiv \int d^3p\, \mathbf{v} f_\alpha(\mathbf{x}, \mathbf{p}, t), \tag{1.3.20}$$

$$n_\alpha(\mathbf{x}, t)\mathbf{P}_\alpha(\mathbf{x}, t) \equiv \int d^3p\, \mathbf{p} f_\alpha(\mathbf{x}, \mathbf{p}, t), \tag{1.3.21}$$

$$\boldsymbol{P}_\alpha(\mathbf{x}, t) \equiv \int d^3p\, [\mathbf{p} - \mathbf{P}_\alpha(\mathbf{x}, t)]\, [\mathbf{v} - \mathbf{V}_\alpha(\mathbf{x}, t)]\, f_\alpha(\mathbf{x}, \mathbf{p}, t), \tag{1.3.22}$$

where $\mathbf{v} = (\mathbf{p}/m_\alpha)(1 + \mathbf{p}^2/m_\alpha^2c^2)^{-1/2}$. Operating on Eq. (1.3.2) with $\int d^3p$, and $\int d^3p\, \mathbf{p}$ gives

$$\frac{\partial}{\partial t} n_\alpha + \nabla \cdot (n_\alpha \mathbf{V}_\alpha) = 0, \tag{1.3.23}$$

$$\frac{\partial}{\partial t}\mathbf{P}_\alpha + \mathbf{V}_\alpha \cdot \nabla \mathbf{P}_\alpha + \frac{\nabla \cdot \boldsymbol{P}_\alpha}{n_\alpha} = e_\alpha \left(\mathbf{E} + \frac{\mathbf{V}_\alpha \times \mathbf{B}}{c} \right). \tag{1.3.24}$$

For a cold plasma, the term $\nabla \cdot \boldsymbol{P}_\alpha / n_\alpha$ is not retained in Eq. (1.3.24). However, if finite temperature effects were to be included in this macroscopic fluid model, this term would be retained and the time development of $\boldsymbol{P}_\alpha(\mathbf{x}, t)$ determined by taking the appropriate momentum moments of Eq. (1.3.2). The chain of moment equations would be closed by making an assumption about the form of the heat flow tensor $Q_\alpha(\mathbf{x}, t)$. Equations (1.3.23) and (1.3.24) are supplemented by the Maxwell equations, that is,

$$\nabla \times \mathbf{E} = -\frac{1}{c}\frac{\partial \mathbf{B}}{\partial t}, \tag{1.3.25}$$

$$\nabla \times \mathbf{B} = \frac{4\pi}{c}\sum_\alpha e_\alpha n_\alpha \mathbf{V}_\alpha + \frac{4\pi}{c}\mathbf{J}_{\text{ext}} + \frac{1}{c}\frac{\partial \mathbf{E}}{\partial t}, \tag{1.3.26}$$

$$\nabla \cdot \mathbf{E} = \sum_\alpha 4\pi n_\alpha e_\alpha + 4\pi\rho_{\text{ext}}, \tag{1.3.27}$$

$$\nabla \cdot \mathbf{B} = 0. \tag{1.3.28}$$

As in the Vlasov description, an equilibrium analysis of Eqs. (1.3.23)-(1.3.28) is carried out by setting $\partial/\partial t = 0$. Dropping the $\nabla \cdot \boldsymbol{P}_\alpha$ term in Eq. (1.3.24) gives

$$\nabla \cdot [n_\alpha^0 \mathbf{V}_\alpha^0] = 0, \tag{1.3.29}$$

$$\mathbf{V}_\alpha^0 \cdot \nabla \mathbf{P}_\alpha^0 = e_\alpha\left(\mathbf{E}^0 + \frac{\mathbf{V}_\alpha^0 \times \mathbf{B}^0}{c}\right), \tag{1.3.30}$$

$$\nabla \times \mathbf{E}^0 = 0, \tag{1.3.31}$$

$$\nabla \times \mathbf{B}^0 = \frac{4\pi}{c}\sum_\alpha e_\alpha n_\alpha^0 \mathbf{V}_\alpha^0 + \frac{4\pi}{c}\mathbf{J}_{\text{ext}}(\mathbf{x}), \tag{1.3.32}$$

$$\nabla \cdot \mathbf{E}^0 = \sum_\alpha 4\pi n_\alpha^0 e_\alpha + 4\pi\rho_{\text{ext}}(\mathbf{x}), \tag{1.3.33}$$

$$\nabla \cdot \mathbf{B}^0 = 0, \tag{1.3.34}$$

where $n_\alpha^0(\mathbf{x})$, $\mathbf{V}_\alpha^0(\mathbf{x})$, $\mathbf{P}_\alpha^0(\mathbf{x})$, $\mathbf{E}^0(\mathbf{x})$, and $\mathbf{B}^0(\mathbf{x})$ are the macroscopic equilibrium quantities. In Chapter 2 equilibrium solutions to Eqs. (1.3.29)-(1.3.34) are examined for a variety of equilibrium configurations.

A stability analysis based on Eqs. (1.3.23)-(1.3.28) proceeds in the following manner. The macroscopic fluid and field quantities are expressed as the sum of their equilibrium values plus a perturbation,

$$\psi(\mathbf{x}, t) = \psi^0(\mathbf{x}) + \delta\psi(\mathbf{x}, t). \tag{1.3.35}$$

Linearization of Eqs. (1.3.23)-(1.3.28) gives

$$\frac{\partial}{\partial t}\delta n_\alpha + \nabla \cdot (n_\alpha^0\, \delta\mathbf{V}_\alpha + \delta n_\alpha \mathbf{V}_\alpha^0) = 0, \tag{1.3.36}$$

$$\frac{\partial}{\partial t}\delta \mathbf{P}_\alpha + \mathbf{V}_\alpha^0 \cdot \nabla \delta \mathbf{P}_\alpha + \delta \mathbf{V}_\alpha \cdot \nabla \mathbf{P}_\alpha^0 \tag{1.3.37}$$

$$= e_\alpha \left(\delta \mathbf{E} + \frac{\mathbf{V}_\alpha^0 \times \delta \mathbf{B}}{c} + \frac{\delta \mathbf{V}_\alpha \times \mathbf{B}^0}{c} \right),$$

$$\nabla \times \delta \mathbf{E} = -\frac{1}{c}\frac{\partial}{\partial t}\delta \mathbf{B}, \tag{1.3.38}$$

$$\nabla \times \delta \mathbf{B} = \frac{4\pi}{c}\sum_\alpha e_\alpha \left(\delta n_\alpha \mathbf{V}_\alpha^0 + n_\alpha^0 \delta \mathbf{V}_\alpha \right) + \frac{1}{c}\frac{\partial}{\partial t}\delta \mathbf{E}, \tag{1.3.39}$$

$$\nabla \cdot \delta \mathbf{E} = 4\pi \sum_\alpha \delta n_\alpha e_\alpha, \tag{1.3.40}$$

$$\nabla \cdot \delta \mathbf{B} = 0. \tag{1.3.41}$$

These equations describe the evolution of the perturbations $\delta n_\alpha(\mathbf{x}, t)$, $\delta \mathbf{V}_\alpha(\mathbf{x}, t)$, $\delta \mathbf{P}_\alpha(\mathbf{x}, t)$, $\delta \mathbf{E}(\mathbf{x}, t)$, and $\delta \mathbf{B}(\mathbf{x}, t)$. The pressure tensor term $\nabla \cdot \boldsymbol{P}_\alpha$ is omitted from Eq. (1.3.37). If these perturbation quantities grow in time or space, the equilibrium is unstable.

In concluding this section it is noted that a cold-plasma description based on Eqs. (1.3.23)-(1.3.28) is equivalent to a Vlasov description provided the distribution function $f_\alpha(\mathbf{x}, \mathbf{p}, t)$ is of the form

$$f_\alpha(\mathbf{x}, \mathbf{p}, t) = n_\alpha(\mathbf{x}, t)\, \delta[\mathbf{p} - \mathbf{P}_\alpha(\mathbf{x}, t)]. \tag{1.3.42}$$

Integration of Eq. (1.3.42) readily gives

$$\int d^3p\, f_\alpha(\mathbf{x}, \mathbf{p}, t) = n_\alpha(\mathbf{x}, t),$$

$$\int d^3p\, \mathbf{p} f_\alpha(\mathbf{x}, \mathbf{p}, t) = n_\alpha(\mathbf{x}, t)\mathbf{P}_\alpha(\mathbf{x}, t),$$

$$\int d^3p\, \mathbf{v} f_\alpha(\mathbf{x}, \mathbf{p}, t) = n_\alpha(\mathbf{x}, t)\mathbf{V}_\alpha(\mathbf{x}, t) \equiv n_\alpha(\mathbf{x}, t)\, \frac{\mathbf{P}_\alpha(\mathbf{x}, t)/m_\alpha}{[1 + \mathbf{P}_\alpha^2(\mathbf{x}, t)/m_\alpha^2 c^2]^{1/2}},$$

and

$$\boldsymbol{P}_\alpha(\mathbf{x}, t) = 0.$$

CHAPTER 2

MACROSCOPIC EQUILIBRIA AND STABILITY

2.1 THE EQUILIBRIUM FORCE EQUATION

In this section the equilibrium properties of a cold, multicomponent, nonneutral plasma column aligned parallel to a uniform external magnetic field, $\mathbf{B}_0^{ext}(\mathbf{x}) = B_0\hat{\mathbf{e}}_z$, as illustrated in Fig. 2.1.1, are considered. The analysis is based on the macroscopic fluid description discussed in Section 1.3.3 [see Eqs. (1.3.29)-(1.3.34)].

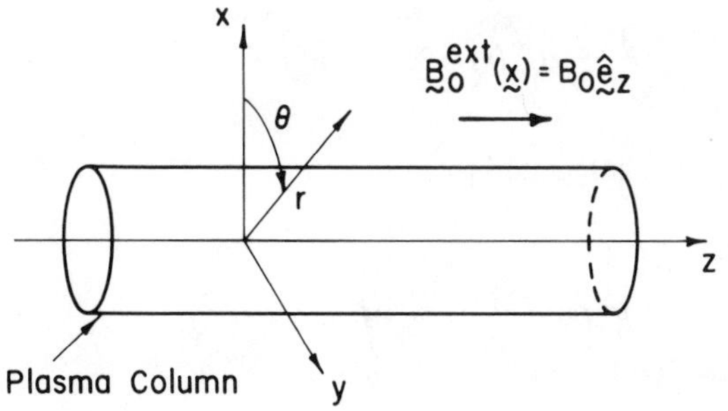

Fig. 2.1.1 Nonneutral plasma column aligned parallel to a uniform external magnetic field, $\mathbf{B}_0^{ext}(\mathbf{x}) = B_0\hat{\mathbf{e}}_z$. Cylindrical polar coordinates (r, θ, z) are introduced with the z-axis coinciding with the axis of symmetry; θ is the polar angle in the x-y plane, and $r = (x^2 + y^2)^{1/2}$ is the radial distance from the z-axis.

The following simplifying assumptions are made regarding the equilibrium configuration:

1. The plasma is infinite and uniform in the z-direction, with $\partial n_\alpha^0(\mathbf{x})/\partial z = 0$ and $\partial \mathbf{V}_\alpha^0(\mathbf{x})/\partial z = 0$, and there is no equilibrium electric field parallel to $B_0\hat{\mathbf{e}}_z$, that is, $\mathbf{E}^0(\mathbf{x}) \cdot \hat{\mathbf{e}}_z = 0$.
2. The equilibrium radial density and velocity profiles are assumed to be azimuthally symmetric about the magnetic axis, that is,

$$n_\alpha^0(\mathbf{x}) = n_\alpha^0(r), \qquad \mathbf{V}_\alpha^0(\mathbf{x}) = \mathbf{V}_\alpha^0(r),$$

where $r = (x^2 + y^2)^{1/2}$ is the radial distance from the axis of symmetry (see Fig. 2.1.1).

In general, the plasma components may have relativistic motion along the guide field $\hat{B}_0\hat{\mathbf{e}}_z$, as well as in the azimuthal direction. Since there is no radial fluid motion in equilibrium, the mean velocity of component α can be expressed as

$$\mathbf{V}_\alpha^0(r) = V_{\alpha\theta}^0(r)\hat{\mathbf{e}}_\theta + V_{\alpha z}^0(r)\hat{\mathbf{e}}_z, \tag{2.1.1}$$

where $\hat{\mathbf{e}}_\theta$ and $\hat{\mathbf{e}}_z$ are unit vectors in the θ- and z-directions, respectively. The equilibrium continuity equation, $\nabla \cdot [n_\alpha^0(r)\,\mathbf{V}_\alpha^0(r)] = 0$, is automatically satisfied for arbitrary $n_\alpha^0(r)$, $V_{\alpha\theta}^0(r)$, and $V_{\alpha z}^0(r)$. For a nonneutral plasma column with azimuthal symmetry, the equilibrium radial electric field, $\mathbf{E}^0(\mathbf{x}) = E_r^0(r)\hat{\mathbf{e}}_r$, is determined from Poisson's equation [see Eq. (1.3.33) with $\rho_{\text{ext}} = 0$]:

$$\frac{1}{r}\frac{\partial}{\partial r} rE_r^0(r) = \sum_\alpha 4\pi e_\alpha n_\alpha^0(r). \tag{2.1.2}$$

Equation (2.1.2) can be integrated to give

$$E_r^0(r) = \frac{4\pi}{r}\sum_\alpha e_\alpha \int_0^r dr'\, r' n_\alpha^0(r'). \tag{2.1.3}$$

It has been assumed that there are no external charge sources in obtaining Eq. (2.1.3). For the equilibrium configuration shown in Fig. 2.1.1, the equilibrium magnetic field $\mathbf{B}^0(\mathbf{x})$ can be expressed as

$$\mathbf{B}^0(\mathbf{x}) = B_0\hat{\mathbf{e}}_z + B_\theta^s(r)\hat{\mathbf{e}}_\theta + B_z^s(r)\hat{\mathbf{e}}_z, \tag{2.1.4}$$

where B_0 is the externally imposed magnetic field, $B_\theta^s(r)$ is the azimuthal self field generated by the equilibrium axial current, $J_z^0(r) = \sum_\alpha n_\alpha^0(r) e_\alpha V_{\alpha z}^0(r)$, and $B_z^s(r)$ is the axial self field generated by the equilibrium azimuthal current,

$J_\theta^0(r) = \Sigma_\alpha n_\alpha^0(r) e_\alpha V_{\alpha\theta}^0(r)$. For example, if no ions are present, and the motions of the electron fluid are in the direction shown in Fig. 2.1.2, the self magnetic

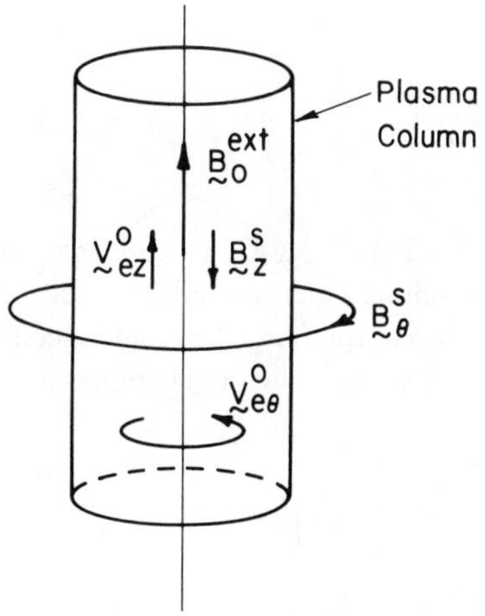

Fig. 2.1.2 Nonneutral plasma column aligned parallel to a uniform external magnetic field, $\mathbf{B}_0^{\text{ext}}(\mathbf{x}) = B_0 \hat{\mathbf{e}}_z$. The axial electron current, $-n_e^0(r) e V_{ez}^0(r) \hat{\mathbf{e}}_z$, and azimuthal electron current, $-n_e^0(r) e V_{e\theta}^0(r) \hat{\mathbf{e}}_\theta$, produce the self magnetic field components, $B_\theta^s(r)\hat{\mathbf{e}}_\theta$ and $B_z^s(r)\hat{\mathbf{e}}_z$, respectively.

fields are in the *negative* z- and θ-directions. The axial self field for this configuration tends to *weaken* the total magnetic field strength in the z-direction. The azimuthal self field (which is strongest at the outer edge of the column) tends to pinch or radially confine the beam. The equilibrium self magnetic field is determined from Eq. (1.3.32) with $\mathbf{J}_{\text{ext}}(\mathbf{x}) = 0$. The two components of this equation are

$$\frac{\partial}{\partial r} B_z^s(r) = -\frac{4\pi}{c} J_\theta^0(r) = -\frac{4\pi}{c} \sum_\alpha e_\alpha n_\alpha^0(r) V_{\alpha\theta}^0(r), \tag{2.1.5}$$

$$\frac{1}{r}\frac{\partial}{\partial r}[r B_\theta^s(r)] = \frac{4\pi}{c} J_z^0(r) = \frac{4\pi}{c} \sum_\alpha e_\alpha n_\alpha^0(r) V_{\alpha z}^0(r). \tag{2.1.6}$$

Equations (2.1.5) and (2.1.6) can be integrated to give

$$B_z^s(r) = \frac{4\pi}{c}\sum_\alpha e_\alpha \int_r^\infty dr'\, n_\alpha^0(r')V_{\alpha\theta}^0(r'), \tag{2.1.7}$$

$$B_\theta^s(r) = \frac{4\pi}{cr}\sum_\alpha e_\alpha \int_0^r dr'\, n_\alpha^0(r')V_{\alpha z}^0(r')r'. \tag{2.1.8}$$

Referring to Eqs. (1.3.29)-(1.3.34), we note that the equilibrium Maxwell equations and the continuity equations are satisfied. The only remaining equation is the radial force equation (the θ- and z-components of the force equation are trivially satisfied). The radial component of Eq. (1.3.30) is

$$-\frac{V_{\alpha\theta}^0 P_{\alpha\theta}^0}{r} = e_\alpha\left[E_r^0 + \frac{V_{\alpha\theta}^0(B_0 + B_z^s)}{c} - \frac{V_{\alpha z}^0 B_\theta^s}{c}\right]. \tag{2.1.9}$$

Since $m_\alpha \mathbf{V}_\alpha^0 = \mathbf{P}_\alpha^0(1 + \mathbf{P}_\alpha^{02}/m_\alpha^2c^2)^{-1/2}$, Eq. (2.1.9) can also be written in the form

$$-\frac{m_\alpha\gamma_\alpha V_{\alpha\theta}^{02}}{r} = e_\alpha\left[E_r^0 + \frac{V_{\alpha\theta}^0(B_0 + B_z^s)}{c} - \frac{V_{\alpha z}^0 B_\theta^s}{c}\right], \tag{2.1.10}$$

where $\gamma_\alpha(r) = (1 - V_{\alpha\theta}^{02}/c^2 - V_{\alpha z}^{02}/c^2)^{-1/2}$. Equation (2.1.10) is a statement of the balance between centrifugal, electrostatic, and magnetic forces on a fluid element.

It is convenient to introduce dimensionless variables for the θ- and z-components of the velocity as follows:

$$\beta_{\alpha\theta}(r) \equiv V_{\alpha\theta}^0(r)/c; \quad \beta_{\alpha z}(r) \equiv V_{\alpha z}^0(r)/c. \tag{2.1.11}$$

If $E_r^0(r)$, $B_z^s(r)$, and $B_\theta^s(r)$ are eliminated from Eq. (2.1.10) by means of Eqs. (2.1.3), (2.1.7), and (2.1.8), Eq. (2.1.10) can be expressed as

$$-\frac{m_\alpha\gamma_\alpha(r)[\beta_{\alpha\theta}(r)c]^2}{r} = e_\alpha\left\{\frac{4\pi}{r}\sum_\eta e_\eta \int_0^r dr'\, r' n_\eta^0(r')\right.$$

(Equation continues on page 21)

$$+\beta_{\alpha\theta}(r)\left[B_0 + 4\pi \sum_\eta e_\eta \int_r^\infty dr'\, \beta_{\eta\theta}(r') n_\eta^0(r')\right]$$

$$-4\pi \frac{\beta_{\alpha z}(r)}{r} \sum_\eta e_\eta \int_0^r dr'\, r' \beta_{\eta z}(r') n_\eta^0(r') \Bigg\}, \tag{2.1.12}$$

where $\gamma_\alpha(r) = (1 - \beta_{\alpha\theta}^2 - \beta_{\alpha z}^2)^{-1/2}$. The term on the left in Eq. (2.1.12) is the centrifugal force on a fluid element, and the first term on the right is the radial electrostatic force on a fluid element. The second term on the right is the $\mathbf{V}_\alpha^0 \times (\mathbf{B}_0^{\mathrm{ext}} + \mathbf{B}_z^s)$ force on a fluid element, and the final term is the $\mathbf{V}_\alpha^0 \times \mathbf{B}_\theta^s$ force on a fluid element.

It is important to note from Eq. (2.1.12) the freedom there is in describing the equilibrium. In particular, any two of the three quantities $n_\alpha^0(r)$, $\beta_{\alpha\theta}(r)$, and $\beta_{\alpha z}(r)$ can be chosen arbitrarily, and then the remaining quantity determined from Eq. (2.1.12) [subject of course to $n_\alpha^0(r) \geqslant 0$ and $\beta_{\alpha\theta}^2 + \beta_{\alpha z}^2 < 1$]. Specific examples are now considered.

2.2 NONRELATIVISTIC NONDIAMAGNETIC EQUILIBRIA

In this section the radial force equation (2.1.12) is used to study the equilibrium properties of nonrelativistic, nondiamagnetic, nonneutral plasma columns. It is assumed that the axial motions of all plasma components are nonrelatvistic, that is,

$$\beta_{\alpha z}^2 \ll 1. \tag{2.2.1}$$

When Eq. (2.2.1) is satisfied, the final term in Eq. (2.1.12) (which corresponds to the $-e_\alpha V_{\alpha z}^0 B_\theta^s/c$ pinching force on a fluid element) can be neglected in comparison with other terms. For example, if no ions are present, the final term in Eq. (2.1.12) is $0(\beta_{ez}^2)$ smaller than the electrostatic force term. It is further assumed that the azimuthal motions are nonrelativistic,

$$\beta_{\alpha\theta}^2 \ll 1, \tag{2.2.2}$$

and that the axial diamagnetic field B_z^s is small compared to the external field B_0 in Eq. (2.1.12),

$$|B_z^s| = |4\pi \sum_\eta e_\eta \int_r^\infty dr' \beta_{\eta\theta}(r')\, n_\eta^0(r')| \ll |B_0|. \tag{2.2.3}$$

The approximation in Eq. (2.2.3) can be justified *a posteriori* and is consistent with the nonrelativistic approximation in Eq. (2.2.2), provided the external magnetic field is sufficiently strong.

Making use of Eqs. (2.2.1)-(2.2.3), and $\gamma_\alpha(r) \simeq 1$, we can write Eq. (2.1.12) in the approximate form

$$-\frac{m_\alpha [\beta_{\alpha\theta}(r)c]^2}{r} = e_\alpha \frac{4\pi}{r} \sum_\eta e_\eta \int_0^r dr'\, r' n_\eta^0(r') + e_\alpha B_0 \beta_{\alpha\theta}(r) \,. \tag{2.2.4}$$

It is convenient to rewrite Eq. (2.2.4), the radial force equation, in terms of the angular velocity $\omega_\alpha(r)$, where

$$\omega_\alpha(r) \equiv \frac{V_{\alpha\theta}^0(r)}{r} = \frac{\beta_{\alpha\theta}(r)c}{r} \,. \tag{2.2.5}$$

Equation (2.2.4) becomes

$$\omega_\alpha^2(r) + \sum_\eta \frac{4\pi e_\alpha e_\eta}{m_\alpha r^2} \int_0^r dr'\, r' n_\eta^0(r') + \epsilon_\alpha \Omega_\alpha \omega_\alpha(r) = 0, \tag{2.2.6}$$

where

$$\Omega_\alpha \equiv \frac{|e_\alpha| B_0}{m_\alpha c}, \quad \text{and} \quad \epsilon_\alpha \equiv \operatorname{sgn} e_\alpha \,. \tag{2.2.7}$$

Equation (2.2.6) can be used in two ways. For example, if $n_\alpha^0(r)$ is specified, Eq. (2.2.6) can be used to determine the rotation rate $\omega_\alpha(r)$, that is,

$$\omega_\alpha(r) = \omega_\alpha^\pm(r) = -\frac{\epsilon_\alpha \Omega_\alpha}{2} \left\{ 1 \pm \left[1 - 4 \sum_\eta \frac{4\pi e_\alpha e_\eta}{m_\alpha \Omega_\alpha^2 r^2} \int_0^r dr'\, r' n_\eta^0(r') \right]^{1/2} \right\} . \tag{2.2.8}$$

Alternatively, if $\omega_\alpha(r)$ is specified, Eq. (2.2.6) can be used to calculate the corresponding density profile $n_\alpha^0(r)$ self-consistently. Multiplying Eq. (2.2.6) by r^2 and differentiating with respect to r gives

$$\frac{4\pi e_\alpha}{m_\alpha} \sum_\eta e_\eta n_\eta^0(r) = -\frac{1}{r}\frac{d}{dr} \left\{ r^2 \left[\omega_\alpha^2(r) + \epsilon_\alpha \Omega_\alpha \omega_\alpha(r) \right] \right\} . \tag{2.2.9}$$

If there are N plasma components, and $\omega_\alpha(r)$ is specified for $\alpha = 1, 2, \ldots, N$, then Eq. (2.2.9) is a system of N equations that determine $n_1^0(r), n_2^0(r), \ldots, n_N^0(r)$. Note from Eq. (2.2.8) that the condition

$$4 \sum_{\eta} \frac{4\pi e_\alpha e_\eta}{m_\alpha \Omega_\alpha^2 r^2} \int_0^r dr'\, r' n_\eta^0(r') \leqslant 1 \tag{2.2.10}$$

must be satisfied if $\omega_\alpha(r)$ is to be real. The restriction imposed by Eq. (2.2.10) is the condition that magnetic restoring forces sufficiently exceed electrostatic forces for radial confinement of the equilibrium.

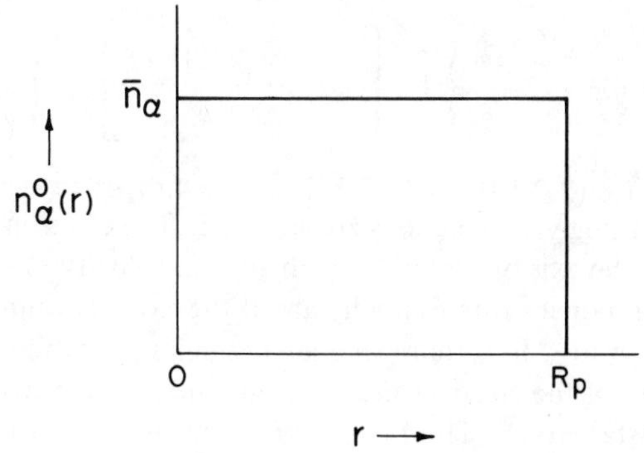

Fig. 2.2.1 Plot of $n_\alpha^0(r)$ versus r for the constant-density profile in Eq. (2.2.11).

Equation (2.2.8) is valid for arbitrary density profiles subject to the inequality in Eq. (2.2.10). As a first example, consider the case in which the density profile is constant for each component (see Fig. 2.2.1), that is,

$$n_\alpha^0(r) = \begin{cases} \bar{n}_\alpha = \text{const.}, & 0 < r < R_p, \\ 0, & r > R_p. \end{cases} \tag{2.2.11}$$

Equation (2.2.8) then reduces to

$$\omega_\alpha(r) = \omega_\alpha^{\pm} = -\frac{\epsilon_\alpha \Omega_\alpha}{2} \left\{ 1 \pm \left[1 - 2 \frac{\sum_\eta 4\pi e_\alpha e_\eta \bar{n}_\eta / m_\alpha}{\Omega_\alpha^2} \right]^{1/2} \right\}, \quad 0 < r < R_p. \tag{2.2.12}$$

If there is a single component of electrons, and no ions are present in the system, Eq. (2.2.12) reduces to the familiar result obtained in Section 1.2 [see Eq. (1.2.5)].† For constant-density profiles, it follows from Eq. (2.2.12) that ω_α = const. in the nonrelativistic, nondiamagnetic regime, that is, there is no shear in angular velocity within the beam. From Eq. (2.2.12), the rotation rates for a constant-density beam consisting of *two* components, electrons and ions with $\bar{n}_i = f\bar{n}_e$, where f = const. = fractional neutralization, are

$$\omega_e = \omega_e^{\pm} = \frac{\Omega_e}{2}\left\{1 \pm \left[1 - \frac{2\bar{\omega}_{pe}^2}{\Omega_e^2}(1-f)\right]^{1/2}\right\}, \tag{2.2.13}$$

$$\omega_i = \omega_i^{\mp} = -\frac{\Omega_i}{2}\left\{1 \pm \left[1 + \frac{2\bar{\omega}_{pi}^2}{\Omega_i^2}\frac{1-f}{f}\right]^{1/2}\right\}, \tag{2.2.14}$$

for $0 < r < R_p$. In Eqs. (2.2.13) and (2.2.14), $\bar{\omega}_{pe}^2 = 4\pi\bar{n}_e e^2/m_e$ and $\bar{\omega}_{pi}^2 = 4\pi\bar{n}_i e^2/m_i$ and singly charged ions are assumed. The electron component rotates rigidly about the axis of symmetry with angular velocity ω_e^+ *or* ω_e^-. Similarly, the ion component rotates rigidly about the axis of symmetry with angular velocity ω_i^+ *or* ω_i^-. In certain parameter regimes the difference between the rotation velocities of the electron and ion components can provide the free energy to drive an instability.[70] This *two-rotating-stream instability* is discussed in Section 2.9.3.

As an example for which there *is* shear in the angular velocity $\omega_\alpha(r)$, consider the hollow density profile illustrated in Fig. 2.2.2. This density profile is expressed as

$$n_\alpha^0(r) = \begin{cases} 0, & 0 < r < R_0, \\ \bar{n}_\alpha = \text{const.}, & R_0 < r < R_p, \\ 0, & r > R_p. \end{cases} \tag{2.2.15}$$

†It is clear from Eq. (2.2.2) that the axial diamagnetic field $B_z^s(r)$ assumes its largest value on axis ($r = 0$). If there is a single component of electrons, and no ions are present in the system, then for a constant-density electron profile (Fig. 2.2.1)

$$\frac{|B_z^s(r=0)|}{|B_0|} = \left|\frac{4\pi e\omega_e R_p^2 \bar{n}_e}{2cB_0}\right| = \left|\frac{\omega_e^2 R_p^2}{c^2}\cdot\frac{\bar{\omega}_{pe}^2/2\Omega_e}{\omega_e}\right|.$$

Since $\beta_{e\theta}^2(R_p) \ll 1$ by assumption [Eq. (2.2.2], and $|\bar{\omega}_{pe}^2/2\Omega_e| \leqslant |\omega_e|$ [see Fig. 1.2.3 and Eq. (1.2.6)], it follows that $|B_z^s(r=0)| \ll |B_0|$, which justifies the neglect of axial diamagnetic effects in Eq. (2.2.4).

Equation (2.2.8) then reduces to

$$\omega_\alpha^\pm(r) = -\frac{\epsilon_\alpha \Omega_\alpha}{2}\left\{1 \pm \left[1 - 2\sum_\eta \frac{4\pi e_\alpha e_\eta \bar{n}_\eta}{m_\alpha \Omega_\alpha^2}\left(1 - \frac{R_0^2}{r^2}\right)\right]^{1/2}\right\},$$

$$R_0 < r < R_p. \qquad (2.2.16)$$

Note from Eq. (2.2.16) that $\partial\omega_\alpha(r)/\partial r \neq 0$ within the beam. As shown in Section 2.10.3, this shear in angular velocity provides the energy source to drive the diocotron instability.[98–111]

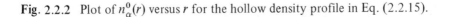

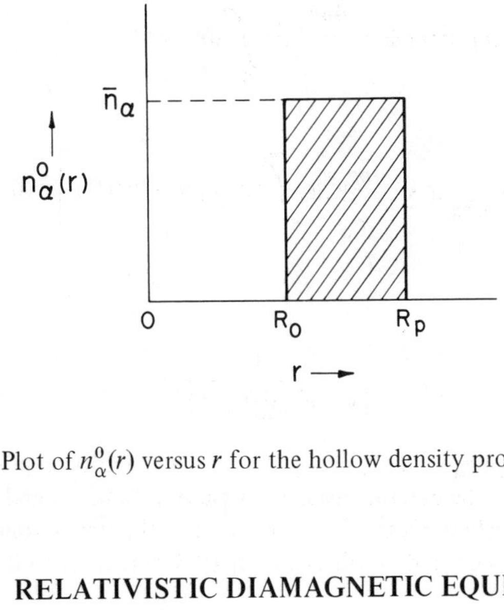

Fig. 2.2.2 Plot of $n_\alpha^0(r)$ versus r for the hollow density profile in Eq. (2.2.15).

2.3 RELATIVISTIC DIAMAGNETIC EQUILIBRIA

In this section properties of the radial force equation, Eq. (2.1.12), are examined for situations in which the mean azimuthal motion of the plasma column is relativistic, and the axial motion is nonrelativistic, that is,

$$\beta_{\alpha z}^2 \ll 1. \qquad (2.3.1)$$

This means that the magnetic pinching force, $-e_\alpha \beta_{\alpha z} B_\theta^s$, is neglected in Eq. (2.1.12), but the radial force produced by the axial diamagnetic field B_z^s is retained in the analysis, where

$$B_z^s(r) = 4\pi \sum_\eta e_\eta \int_r^\infty dr' \, \beta_{\eta\theta}(r') n_\eta^0(r'). \tag{2.3.2}$$

It is also assumed that

$$\beta_{\alpha z}^2 \ll 1 - \beta_{\alpha\theta}^2. \tag{2.3.3}$$

Introducing the angular velocity $\omega_\alpha(r) = V_{\alpha\theta}^0(r)/r = \beta_{\alpha\theta}(r)c/r$, we can express the radial force equation (2.1.12) in the approximate form

$$\gamma_\alpha(r)\omega_\alpha^2(r) + \sum_\eta \frac{4\pi e_\alpha e_\eta}{m_\alpha r^2} \int_0^r dr' \, r' n_\eta^0(r')$$

$$+ \omega_\alpha(r) \left[\Omega_\alpha \epsilon_\alpha + \sum_\eta \frac{4\pi e_\alpha e_\eta}{m_\alpha c^2} \int_r^\infty dr' \, \omega_\eta(r') r' n_\eta^0(r') \right] = 0, \tag{2.3.4}$$

where

$$\gamma_\alpha(r) = \frac{1}{[1 - r^2 \omega_\alpha^2(r)/c^2]^{1/2}}. \tag{2.3.5}$$

In Eq. (2.3.4), $\epsilon_\alpha = \text{sgn}\, e_\alpha$, and $\Omega_\alpha = |e_\alpha| B_0/m_\alpha c$ is the αth-component cyclotron frequency in the external magnetic field B_0. In the nonrelativistic nondiamagnetic limit where $r^2 \omega_\alpha^2/c^2 \ll 1$, $\gamma_\alpha \simeq 1$ and the final term on the left of Eq. (2.3.4) can be neglected. In this case Eq. (2.3.4) reduces to Eq. (2.2.6), as expected.

As in the nonrelativistic nondiamagnetic regime, two points of view can be taken in analyzing the equilibrium force relation, Eq. (2.3.4). First, if the density profiles $n_\alpha^0(r)$ are known, the mean rotational velocity $\omega_\alpha(r)$ can be calculated for each plasma component. Alternatively, if $\omega_\alpha(r)$ is specified for each plasma component, the corresponding density profiles $n_\alpha^0(r)$ can be calculated self-consistently from Eq. (2.3.4). Because of the complicated structure of Eq. (2.3.4), introduced by the factor γ_α, the latter approach is simpler.

For present purposes it is instructive to study the equilibrium force equation (2.3.4) for a one-component pure electron gas column with no ions present in the system, that is,

$$n_i^0(r) = 0. \tag{2.3.6}$$

The equilibrium force equation for the electrons can be expressed as

$$0 = \gamma_e(r)\omega_e^2(r) + \frac{1}{r^2}\int_0^r dr'\, r'\omega_{pe}^2(r')$$

$$-\omega_e(r)\left[\Omega_e - \frac{1}{c^2}\int_r^\infty dr'\, \omega_e(r')r'\omega_{pe}^2(r')\right], \tag{2.3.7}$$

where $\omega_{pe}^2(r) = 4\pi n_e^0(r)e^2/m_e$. Equation (2.3.7) has been analyzed by Bogema,[138] and the general solution for $\omega_{pe}^2(r)$ determined for arbitrary velocity profile $\omega_e(r)$. The *cold-fluid* equilibrium described by Eq. (2.3.7) cannot support solutions for which the total axial magnetic field reverses direction $[B_0 + B_z^s(r) < 0]$. If there is field reversal in the column interior, the radius r_0 at which the total axial magnetic field passes through zero satisfies $B_0 + B_z^s(r_0) = 0$. This implies that the term in square brackets in Eq. (2.3.7) is identically zero at $r = r_0$, that is,

$$\Omega_e = \frac{1}{c^2}\int_{r_0}^\infty dr'\, \omega_e(r')r'\omega_{pe}^2(r'). \tag{2.3.8}$$

Since the remaining terms in Eq. (2.3.7) are nonnegative at $r = r_0$, we conclude that Eqs. (2.3.7) and (2.3.8) are consistent *only if* $\omega_e(r_0) = 0$, and $\omega_{pe}^2(r) = 0$, for $r \leqslant r_0$. For a hollow density profile with $\omega_{pe}^2(r \leqslant r_0) = 0$, and an appropriately chosen velocity profile $\omega_e(r)$ consistent with Eq. (2.3.7), diamagnetic effects can (in principle) reduce the total axial magnetic field to zero at $r = r_0$.

As a first application of Eq. (2.3.7), consider the case in which

$$\gamma_e(r)\omega_e(r) \equiv \omega_0 = \text{const.} \tag{2.3.9}$$

Equation (2.3.9) corresponds to an equilibrium with $P_{e\theta}^0(r)/r = \text{const.}$ for each fluid element. The outer radius of the plasma column is taken to be R_p, that is,

$$n_e^0(r) = 0, \quad r > R_p. \tag{2.3.10}$$

Solving Eq. (2.3.9) for $\omega_e(r)$ gives

$$\omega_e(r) = \frac{\omega_0}{(1 + r^2\omega_0^2/c^2)^{1/2}}, \quad 0 < r < R_p. \tag{2.3.11}$$

Evidently, the angular velocity of mean rotation decreases monotonically from $\omega_e = \omega_0$, for $r^2\omega_0^2/c^2 \ll 1$, to $\omega_e = \omega_0(1 + R_p^2\omega_0^2/c^2)^{-1/2}$, for $r = R_p$. Since $\omega_e(r)$ is specified [Eq. (2.3.11)], Eq. (2.3.7) is an integral equation for $\omega_{pe}^2(r)$. The solution to Eq. (2.3.7), consistent with Eqs. (2.3.10) and (2.3.11), is[138]

$$\omega_{pe}^2(r) = \begin{cases} \omega_{pe}^2(0)\left(1 + \dfrac{r^2\omega_0^2}{2c^2}\right)\exp\left(\dfrac{r^2\omega_0^2}{2c^2}\right), & 0 < r < R_p, \\ 0, & r > R_p. \end{cases} \quad (2.3.12)$$

In Eq. (2.3.12), $\omega_{pe}^2(0) = 4\pi n_e^0(r = 0)e^2/m_e$ is related to Ω_e, ω_0 and R_p by

$$\frac{\omega_{pe}^2(0)}{2} = (\omega_0\Omega_e - \omega_0^2)\,\frac{\exp(-R_p^2\omega_0^2/2c^2)}{[1 + (R_p^2\omega_0^2/c^2)]^{1/2}}\,. \quad (2.3.13)$$

Note from Eqs. (2.3.12) and (2.3.13) that $0 < \omega_0 < \Omega_e$ is required for $\omega_{pe}^2(r) \geqslant 0$.

In the case where the azimuthal motion is nonrelativistic, $R_p^2\omega_0^2/c^2 \ll 1$, Eq. (2.3.12) corresponds to a rectangular density profile with $\omega_{pe}^2(r) \simeq \omega_{pe}^2(0)$ for $0 < r < R_p$(Fig. 2.3.1). Furthermore, for $R_p^2\omega_0^2/c^2 \ll 1$, Eqs. (2.3.11) and (2.3.13) give $\omega_e \simeq \omega_0 \simeq \omega_e^{\pm}$, where $\omega_e^{\pm}$ are the rotor frequencies defined in Eq. (2.2.13) (for $f = 0$).

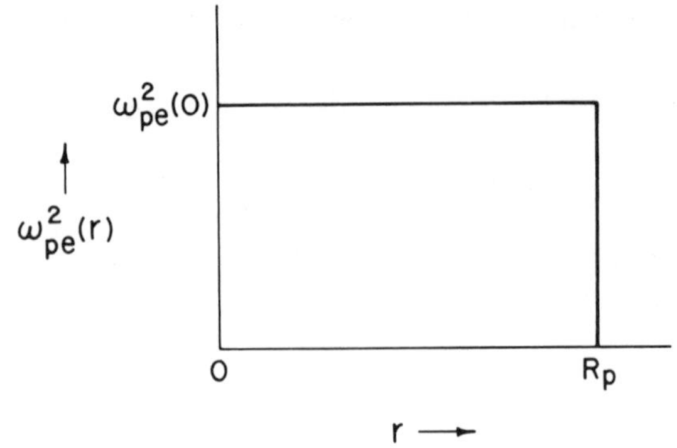

Fig. 2.3.1 Plot of $\omega_{pe}^2(r)$ versus r for $R_p^2\omega_0^2/c^2 \ll 1$ [Eq. (2.3.12)].

In the case where the azimuthal motion near the outer perimeter of the plasma column is relativistic, Eq. (2.3.12) gives a density profile peaked at $r = R_p$.

For example, if $R_p^2\omega_0^2/c^2 = 2$,† then $\omega_{pe}^2(R_p) \simeq 2\omega_{pe}^2(0) \times 2.718$, and the density profile in Eq. (2.3.12) has the form illustrated in Fig. 2.3.2. Evidently the ratio $\omega_{pe}^2(R_p)/\omega_{pe}^2(0)$ can be made arbitrarily large by appropriate choice of $R_p\omega_0/c$. Physically, two effects cause the density to peak off axis in the relativistic case. First, the increase in centrifugal force causes more electrons to be in equilibrium at larger radii. Second, diamagnetic effects weaken the magnetic restoring forces in the column interior. To illustrate the magnitude of the diamagnetic effect for the equilibrium described by Eqs. (2.3.11)-(2.3.13), the quantity $[B_0 + B_z^s(r)]/B_0$, is calculated, that is,

$$\frac{B_0 + B_z^s(r)}{B_0} = 1 - \frac{\omega_{pe}^2(0)}{\omega_0\Omega_e}\int_{r\omega_0/c}^{R_p\omega_0/c} dx\,\frac{x}{(1+x^2)^{1/2}}(1+x^2/2)\exp(x^2/2), \tag{2.3.14}$$

where $\omega_{pe}^2(0)$, ω_0, Ω_e, and R_p are related by Eq. (2.3.13). In Fig. 2.3.3, $[B_0 + B_z^s(r)]/B_0$ is plotted versus r for $R_p^2\omega_0^2/c^2 = 2$ (the ratio used in Fig. 2.3.2), and $\omega_0/\Omega_e = 0.05$. For this choice of parameters, $\omega_{pe}^2(0)/\omega_0\Omega_e = 0.404$ follows from Eq. (2.3.13), and the diamagnetic field B_z^s produces a 75% depression in the axial magnetic field at $r = 0$.

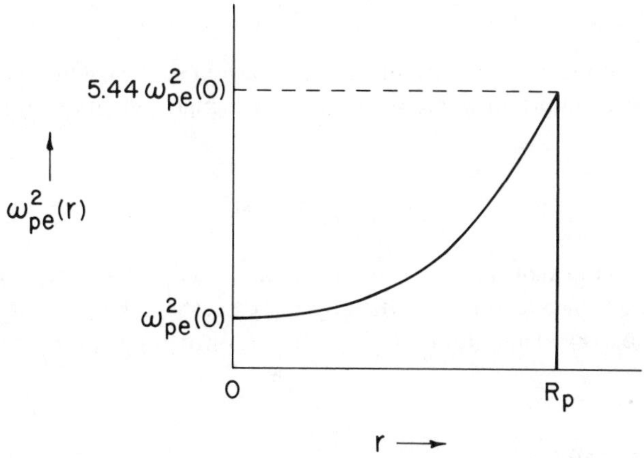

Fig. 2.3.2 Plot of $\omega_{pe}^2(r)$ versus r for $R_p^2\omega_0^2/c^2 = 2$[Eq. (2.3.12)].

---

†Note from Eq. (2.3.11) that $R_p^2\omega_e^2(R_p)/c^2 < 1$ for any choice of $R_p^2\omega_0^2/c^2$.

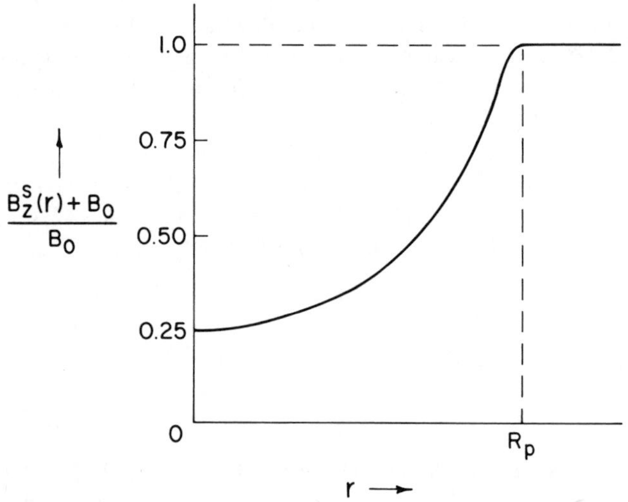

Fig. 2.3.3 Plot of total axial magnetic field, $B_0 + B_z^s(r)$, versus r for $R_p^2\omega_0^2/c^2 = 2$ (the ratio used in Fig. 2.3.2) and $\omega_0/\Omega_e = 0.05$ [Eq. (2.3.14)].

As a second application of the radial force equation (2.3.7), consider the case of a *rigid-rotor* equilibrium where the rotational angular velocity $\omega_e(r)$ is constant,

$$\omega_e(r) = \omega_0 = \text{const.} \tag{2.3.15}$$

As before, the outer radius of the plasma column is taken to be R_p, with $n_e^0(r) = 0$ for $r > R_p$. Since $\omega_e(r)$ is specified by Eq. (2.3.15), Eq. (2.3.7) is an integral equation for $\omega_{pe}^2(r)$. The solution to Eq. (2.3.7) for $\omega_e(r) = \omega_0 = $ const. is[138]

$$\omega_{pe}^2(r) = \begin{cases} \dfrac{\omega_{pe}^2(0)}{(1 - r^2\omega_0^2/c^2)^2}\left[1 + \dfrac{2\omega_0^2}{\omega_{pe}^2(0)} - \dfrac{2\omega_0^2}{\omega_{pe}^2(0)}\dfrac{1 + r^2\omega_0^2/2c^2}{(1 - r^2\omega_0^2/c^2)^{1/2}}\right], & 0 < r < R_p, \\ 0, & r > R_p, \end{cases} \tag{2.3.16}$$

where $\omega_{pe}^2(0) = 4\pi n_e^0(r=0)e^2/m_e$, and $\omega_{pe}(0)$, Ω_e, ω_0, and R_p are related by

$$\frac{\omega_{pe}^2(0)}{2} = (\omega_0\Omega_e - \omega_0^2)\left\{1 - \frac{R_p^2\omega_0^2/c^2}{\omega_0\Omega_e - \omega_0^2}\left[\omega_0\Omega_e - \frac{\omega_0^2}{(1-R_p^2\omega_0^2/c^2)^{1/2}}\right]\right\}. \tag{2.3.17}$$

It is straightforward to verify from Eqs. (2.3.16) and (2.3.17) that $0 < \omega_0(1 - R_p^2\omega_0^2/2c^2) < \Omega_e(1 - R_p^2\omega_0^2/c^2)^{3/2}$ is required for $\omega_{pe}^2(r) \geqslant 0$ over the range $0 < r < R_p$. In the case where the azimuthal motion is nonrelativistic, $R_p^2\omega_0^2/c^2 \ll 1$, Eq. (2.3.16) corresponds to a rectangular density profile with $\omega_{pe}^2(r) \simeq \omega_{pe}^2(0)$ for $0 < r < R_p$ (Fig. 2.3.1). Furthermore, for $R_p^2\omega_0^2/c^2 \ll 1$, Eq. (2.3.17) gives $\omega_0 \simeq \omega_e^\pm$, where $\omega_e^\pm$ are the rotor frequencies defined in Eq. (2.2.13) (for $f = 0$).

In the case where the azimuthal motion near the outer perimeter of the column is relativistic, the detailed form of the density profile given in Eq. (2.3.16) depends on the value of the parameter δ, where

$$\delta \equiv \frac{\omega_{pe}^2(0) + 2\omega_0^2}{3\omega_0^2}. \tag{2.3.18}$$

For $\delta > 1$, it can be shown from Eq. (2.3.16) that $\omega_{pe}^2(r)$ *increases* away from the axis ($r = 0$), reaches a maximum at some $r = R_0 < R_p$, and decreases to its value $\omega_{pe}^2(R_p)$ at the boundary. For $2/3 < \delta < 1$, however, $\omega_{pe}^2(r)$ decreases monotonically as a function of r. As an example, the density profile given by Eq. (2.3.16) is plotted in Fig. 2.3.4 for $\delta = 4/3$ and $R_p^2\omega_0^2/c^2 = 0.64$. From Eq. (2.3.17), the corresponding value of rigid-rotor frequency is $\omega_0/\Omega_e = 0.386$.

An interesting subclass of density profiles described by Eq. (2.3.16) is the one in which $\omega_{pe}^2(r)$ approaches *zero* (continuously) as $r \to R_p$. Enforcing the condition $\omega_{pe}^2(R_p) = 0$ in Eq. (2.3.16) gives

$$\frac{\omega_{pe}^2(0)}{2\omega_0^2} = \frac{1 + R_p^2\omega_0^2/2c^2}{(1 - R_p^2\omega_0^2/c^2)^{1/2}} - 1, \tag{2.3.19}$$

which relates $\omega_{pe}(0)$, ω_0, and R_p. Combining Eqs. (2.3.17) and (2.3.19) gives

$$\frac{\omega_0}{\Omega_e} = \frac{(1 - R_p^2\omega_0^2/c^2)^{3/2}}{1 - R_p^2\omega_0^2/2c^2}, \tag{2.3.20}$$

which relates ω_0, R_p, and Ω_e. The density profile consistent with Eqs. (2.3.16) and (2.3.19) is[138, 139]

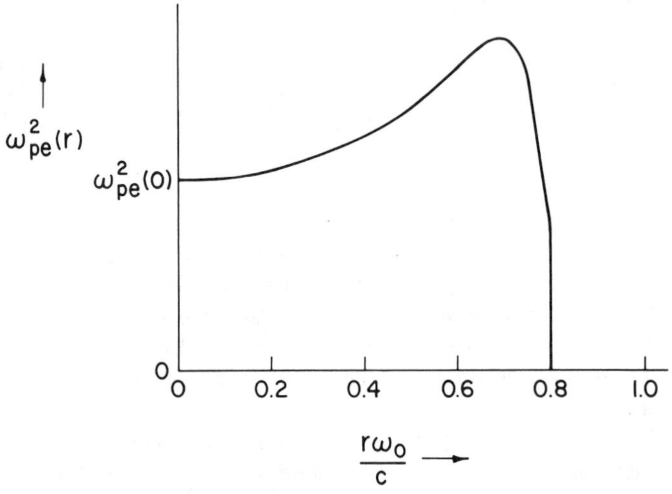

Fig. 2.3.4 Plot of $\omega_{pe}^2(r)$ versus $r\omega_0/c$ for $R_p^2\omega_0^2/c^2 = 0.64$, $\delta = 4/3$, and $\omega_0/\Omega_e = 0.386$ [Eq. (2.3.16)].

$$\omega_{pe}^2(r) = \begin{cases} \dfrac{2\omega_0^2}{(1-r^2\omega_0^2/c^2)^2}\left[\dfrac{1+R_p^2\omega_0^2/2c^2}{(1-R_p^2\omega_0^2/c^2)^{1/2}} - \dfrac{1+r^2\omega_0^2/2c^2}{(1-r^2\omega_0^2/c^2)^{1/2}}\right], & 0<r<R_p, \\ 0, & r>R_p. \end{cases} \tag{2.3.21}$$

In Fig. 2.3.5 the density profile given by Eq. (2.3.21) is plotted versus $r\omega_0/c$, for $R_p\omega_0/c = 0.7$ and $R_p\omega_0/c = 0.9$. To illustrate the diamagnetic effect for the equilibrium described by Eqs. (2.3.19)-(2.3.21), $[B_0 + B_z^s(r)]/B_0$ is plotted versus $r\omega_0/c$ in Fig. 2.3.6 for the parameters chosen in Fig. 2.3.5.

2.4 RELATIVISTIC ELECTRON BEAM EQUILIBRIA

In this section the radial force equation (2.1.12) is used to study the equilibrium properties of a relativistic electron beam propagating parallel to a uniform, externally applied guide field $B_0\hat{\mathbf{e}}_z$. The positive ions are taken to form a stationary, infinitely massive background which is at rest in the laboratory frame.

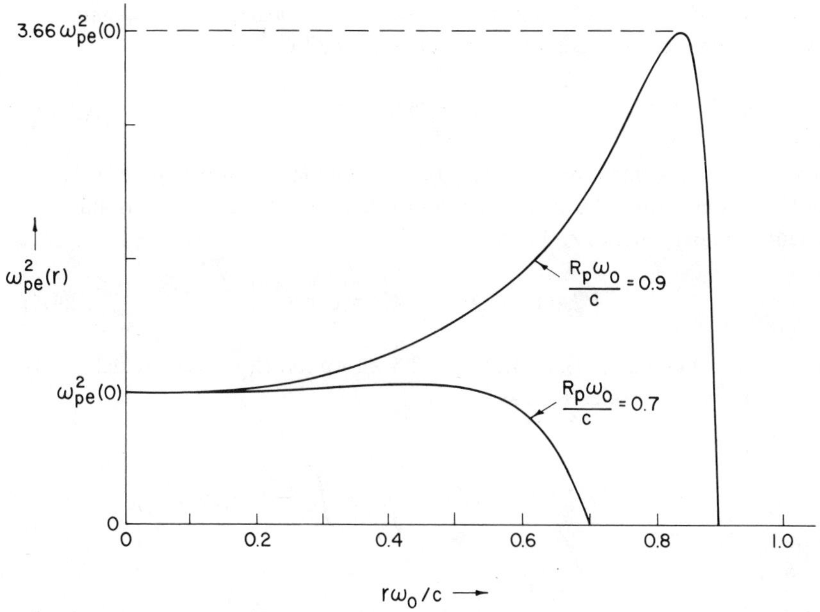

Fig. 2.3.5 Plots of $\omega_{pe}^2(r)$ versus $r\omega_0/c$ for $R_p\omega_0/c = 0.7$ and $R_p\omega_0/c = 0.9$ [Eq. (2.3.21)].

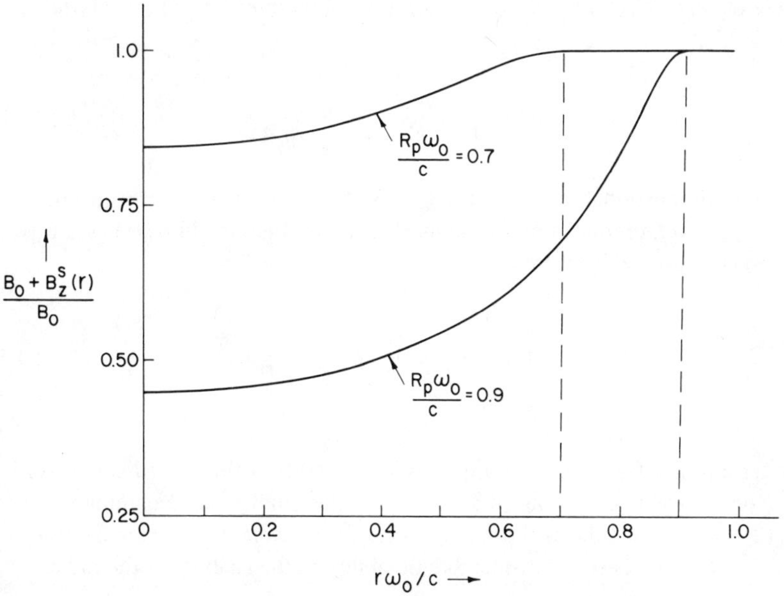

Fig. 2.3.6 Plots of total axial magnetic field, $B_0 + B_z^s(r)$, versus $r\omega_0/c$ for the equilibrium parameters chosen in Fig. 2.3.5.

The ions provide partial neturalization of the electron beam, and it is assumed that the ion and electron density profiles are related by [95,96]

$$n_i^0(r) = f n_e^0(r), \tag{2.4.1}$$

where f = const. = fractional neutralization (single ionization is assumed). It is further assumed that the axial velocity profile of the electron beam is independent of radial distance r, that is,

$$V_{ez}^0(r) = \beta_{ez}(r)c = \beta_0 c = \text{const.} \tag{2.4.2}$$

Making use of Eqs. (2.1.12), (2.4.1) and (2.4.2), we can express the radial force equation for the electrons as

$$0 = \gamma_e(r)\omega_e^2(r) + \frac{1}{r^2}(1 - f - \beta_0^2)\int_0^r dr'\, r'\omega_{pe}^2(r')$$

$$-\omega_e(r)\left[\Omega_e - \frac{1}{c^2}\int_r^\infty dr'\, \omega_e(r')r'\omega_{pe}^2(r')\right], \tag{2.4.3}$$

where $\omega_e(r) = V_{e\theta}^0(r)/r = \beta_{e\theta}(r)c/r$, $\omega_{pe}^2(r) = 4\pi n_e^0(r)e^2/m_e$, $\Omega_e = eB_0/m_e c$, and

$$\gamma_e(r) = \frac{1}{[1 - \beta_0^2 - r^2\omega_e^2(r)/c^2]^{1/2}}. \tag{2.4.4}$$

The term proportional to $-\beta_0^2$ in Eq. (2.4.3) is a result of the *inward* magnetic force, $e\beta_0 B_\theta^s(r)$, produced by the azimuthal self magnetic field $B_\theta^s(r)$ [see Eqs. (2.1.8) and (2.1.12)], where

$$B_\theta^s(r) = -\frac{m_e}{e}\frac{\beta_0}{r}\int_0^r dr'\, r'\omega_{pe}^2(r'). \tag{2.4.5}$$

The term proportional to $-f$ in Eq. (2.4.3) is a result of the *inward* electrostatic force on the electrons produced by the ion background. The remaining terms in Eq. (2.4.3) are the same as those in Eq. (2.3.7) and were discussed in Section 2.3.† As before, there is considerable flexibility in the analysis of the radial

†Note that in the limit where there are no ions ($f = 0$), and the axial motion is nonrelativistic with $\beta_0^2 \ll 1 - \beta_{e\theta}^2(r)$, Eq. (2.4.3) reduces to Eq. (2.3.7), as expected.

force equation. For example, if f, β_0^2, and $\omega_{pe}^2(r)$ are specified, Eq. (2.4.3) can be used to determine the angular velocity profile $\omega_e(r)$ self-consistently.

As a first application of Eq. (2.4.3) it is informative to ascertain whether the radial force equation supports equilibrium solutions for which there is *no* rotation of the electron beam, that is,

$$\omega_e(r) = 0. \tag{2.4.6}$$

In this case Eq. (2.4.3) is satisfied for *arbitrary* density profile $\omega_{pe}^2(r)$ provided

$$1 - f = \beta_0^2 . \tag{2.4.7}$$

Equation (2.4.7) is a statement that the outward electrostatic force (which is proportional to $1-f$) must *exactly* balance the inward magnetic pinching force (which is proportional to β_0^2) in order for a nonrotating, cold-fluid equilibrium to exist. Equation (2.4.7) places a severe restriction on values of f and β_0^2. In Section 2.5 a simple macroscopic model of a nonrotating Bennett pinch[44] is constructed for the case $\beta_0^2 > 1-f$. In this model, the difference between the magnetic pinching force and the electrostatic repulsive force is compensated by a radial pressure gradient.

As a second application of the radial force equation (2.4.3), the restriction $\omega_e(r) = 0$ is removed but the mean azimuthal motion of the electron beam is assumed to be *nonrelativistic* with

$$\frac{r^2\,\omega_e^2(r)}{c^2} = \beta_{e\theta}^2(r) \ll 1 - \beta_0^2 \,. \tag{2.4.8}$$

In this case the axial diamagnetic field contribution [the final term in Eq. (2.4.3)] can be neglected, and $\gamma_e(r)$ can be approximated by

$$\gamma_e(r) = \frac{1}{(1-\beta_0^2)^{1/2}} \equiv \gamma_0 . \tag{2.4.9}$$

Making use of Eqs. (2.4.8) and (2.4.9), we can express Eq. (2.4.3) in the approximate form

$$0 = \gamma_0\omega_e^2(r) + \frac{1}{r^2}(1-f-\beta_0^2)\int_0^r dr'\, r'\omega_{pe}^2(r') - \omega_e(r)\Omega_e . \tag{2.4.10}$$

Solving Eq. (2.4.10) for the angular velocity profile $\omega_e(r)$ gives

$$\omega_e(r)=\omega_e^{\pm}(r)=\frac{\Omega_e}{2\gamma_0}\left\{1\pm\left[1-\frac{4\gamma_0(1-f-\beta_0^2)}{\Omega_e^2 r^2}\int_0^r dr'\,r'\omega_{pe}^2(r')\right]^{1/2}\right\},\tag{2.4.11}$$

where $\Omega_e = eB_0/m_e c$. Note that Eq. (2.4.11) reduces to Eq. (2.2.8) for $\alpha = e$, $\beta_0^2 \ll 1-f$, and $\gamma_0 \simeq 1$.

In the case where there is no external guide field ($\Omega_e = 0$), the condition for the radical in Eq. (2.4.11) to be real is

$$\beta_0^2 \geqslant 1-f,\tag{2.4.12}$$

and the angular velocity of mean rotation is [see Eq. (2.4.10)]

$$\omega_e(r)=\omega_e^{\pm}(r)=\frac{1}{r}\left[\frac{\beta_0^2-1+f}{\gamma_0}\int_0^r dr'\,r'\omega_{pe}^2(r')\right]^{1/2}.\tag{2.4.13}$$

Equation (2.4.12) is a statement that the magnetic pinching force must be at least as large as the electrostatic repulsive force in order for the equilibrium to exist. If $\beta_0^2 = 1-f$, then $\omega_e(r) = 0$. If $\beta_0^2 > 1-f$, in the present cold-plasma model the beam rotates with mean angular velocity given by Eq. (2.4.13).

In the case where there is an external guide field ($\Omega_e \neq 0$), the condition for radial confinement of the beam is not as stringent as Eq. (2.4.12) since the external magnetic field B_0 provides an additional restoring force in the radial direction. For example, for a constant-density profile (see Fig. 2.4.1)

$$\omega^2{}_{pe}(r)=\begin{cases}\dfrac{4\pi\bar{n}_e e^2}{m_e}=\overline{\omega}_{pe}^2, & 0<r<R_p,\\[2ex] 0, & r>R_p,\end{cases}\tag{2.4.14}$$

Eq. (2.4.11) reduces to

$$\omega_e(r)=\omega_e^{\pm}(r)=\frac{\Omega_e}{2\gamma_0}\left\{1\pm\left[1-\frac{2\gamma_0\overline{\omega}_{pe}^2}{\Omega_e^2}(1-f-\beta_0^2)\right]^{1/2}\right\},\quad 0<r<R_p.\tag{2.4.15}$$

The condition for the radical in Eq. (2.4.15) to be real is

$$\beta_0^2+\Omega_e^2/2\gamma_0\overline{\omega}_{pe}^2 \geqslant 1-f.\tag{2.4.16}$$

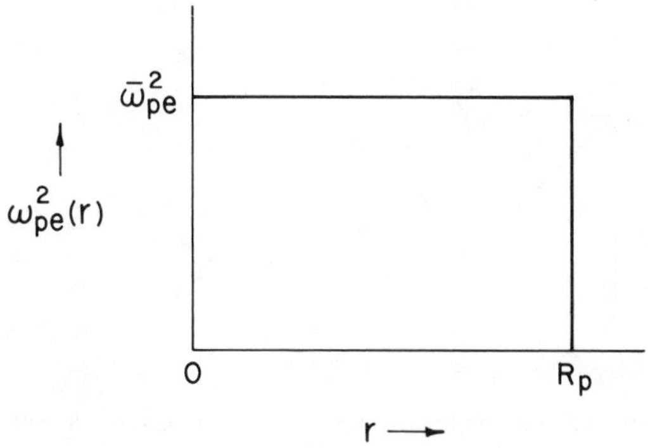

Fig. 2.4.1 Plot of $\omega_{pe}^2(r)$ versus r for the constant-density profile in Eq. (2.4.14).

In the absence of a guide field, Eq. (2.4.16) reduces to Eq. (2.4.12), and a partially neutralizing ion background must be present ($f \neq 0$) in order for the electron beam to propagate.[95,96] (Keep in mind that $\beta_0^2 < 1$ is required.) If an external guide field is present, however, Eq. (2.4.16) implies that the equilibrium exists even if there is *no* ion background ($f = 0$), as long as the inequality $\beta_0^2 \geqslant 1 - \Omega_e^2/2\gamma_0\bar{\omega}_{pe}^2$ is satisfied. For the constant-density and constant-angular-velocity profiles given by Eqs. (2.4.14) and (2.4.15), the self-consistent azimuthal magnetic field $B_\theta^s(r)$ generated by the axial electron current is [see Eq. (2.4.5) and Fig. 2.4.2]

$$B_\theta^s(r) = \begin{cases} -\dfrac{m_e\beta_0\bar{\omega}_{pe}^2}{2e}\, r, & 0 < r < R_p, \\[2ex] -\dfrac{m_e\beta_0\bar{\omega}_{pe}^2}{2e}\,\dfrac{R_p^2}{r}, & r > R_p. \end{cases} \tag{2.4.17}$$

The *magnitude* of the azimuthal self magnetic field increases linearly from zero (at $r = 0$) to a maximum value of $m_e\beta_0\bar{\omega}_{pe}^2 R_p/2e$ at the outer edge of the beam ($r = R_p$). Outside of the beam ($r > R_p$), the magnitude of the azimuthal self magnetic field exhibits a $1/r$ dependence.

As a further example that utilizes Eq. (2.4.11), consider the case in which the electron density profile is hollow (see Fig. 2.2.2) with

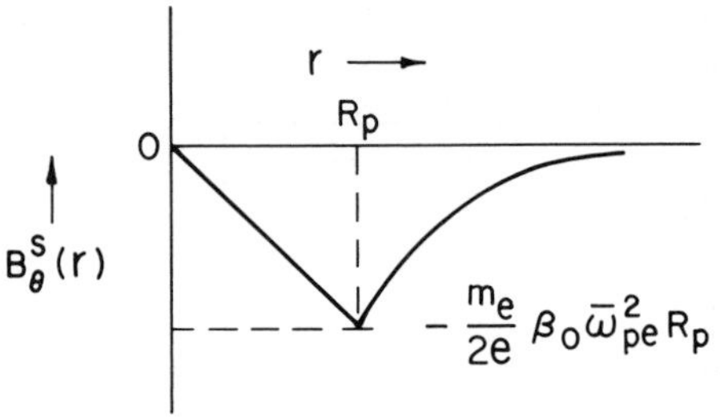

Fig. 2.4.2 Plot of azimuthal self magnetic field $B_\theta^s(r)$ versus r for a constant-density electron beam [Eq. (2.4.17)].

$$\omega_{pe}^2(r) = \begin{cases} 0, & 0 < r < R_0, \\ \dfrac{4\pi \bar{n}_e e^2}{m_e} = \bar{\omega}_{pe}^2, & R_0 < r < R_p, \\ 0, & r > R_p. \end{cases} \tag{2.4.18}$$

Substituting Eq. (2.4.18) into Eq. (2.4.11) gives

$$\omega_e(r) = \omega_e^{\pm}(r) = \frac{\Omega_e}{2\gamma_0}\left\{1 \pm \left[1 - \frac{2\gamma_0 \bar{\omega}_{pe}^2}{\Omega_e^2}(1 - f - \beta_0^2)\left(1 - \frac{R_0^2}{r^2}\right)\right]^{1/2}\right\},$$

$$R_0 < r < R_p. \tag{2.4.19}$$

As in the case where the axial motion is nonrelativistic (see Section 2.2), the hollow beam profile [Eq. (2.4.18)] has a corresponding shear in angular velocity, that is, $\partial\omega_e(r)/\partial r \neq 0$ in Eq. (2.4.19).

For the density and angular velocity profiles given by Eqs. (2.4.18) and (2.4.19), the self-consistent azimuthal magnetic field $B_\theta^s(r)$ generated by the axial electron current is [see Eq. (2.4.5)]

$$B_\theta^s(r) = \begin{cases} 0, & 0 < r < R_0, \\ -\dfrac{m_e \beta_0 \bar{\omega}_{pe}^2}{2e}\dfrac{r^2 - R_0^2}{r}, & R_0 < r < R_p, \\ -\dfrac{m_e \beta_0 \bar{\omega}_{pe}^2}{2e}\dfrac{(R_p^2 - R_0^2)}{r}, & r > R_p. \end{cases} \tag{2.4.20}$$

As a final application of the radial force equation (2.4.3), the restriction that the mean azimuthal motion of the electron beam be nonrelativisitc [Eq. (2.4.8)] is removed. In this case there is no simplification in the form of the radial force equation (2.4.3), or in the definition of $\gamma_e(r)$ given in Eq. (2.4.4). However, since Eq. (2.4.3) is similar in structure† to the equilibrium equation (2.3.7), many of the techniques used in solving Eq. (2.3.7) can be applied to Eq. (2.4.3). For example, if the angular velocity profile $\omega_e(r)$ is specified, the density profile $\omega_{pe}^2(r)$ can be determined self-consistently from Eq. (2.4.3), using the general procedure discussed by Bogema.[138] As a specific example, consider the case of a *rigid-rotor* equilibrium where the angular velocity $\omega_e(r)$ is constant, that is,

$$\omega_e(r) = \omega_0 = \text{const.} \tag{2.4.21}$$

The outer radius of the electron beam is taken to be R_p with $\omega_{pe}^2(r) = 0$ for $r > R_p$. Since $\omega_e(r)$ is specified by Eq. (2.4.21), Eq. (2.4.3) is an integral equation for $\omega_{pe}^2(r)$. The solution to Eq. (2.4.3) for $\omega_e(r) = \omega_0 = \text{const.}$ is

$$\omega_{pe}^2(r) = \frac{\omega_{pe}^2(0)}{(1-\beta_0^2 - r^2\omega_0^2/c^2 - f)^2}\left\{(1-\beta_0^2-f)^2 + \frac{2\omega_0^2(1-\beta_0^2-f)}{\omega_{pe}^2(0)(1-\beta_0^2)^{1/2}}\right.$$
$$\left. - \frac{2\omega_0^2}{\omega_{pe}^2(0)}\left[\frac{(1-\beta_0^2-f) + \dfrac{r^2\omega_0^2}{2c^2}\,\dfrac{1-\beta_0^2-r^2\omega_0^2/c^2-f}{1-\beta_0^2-r^2\omega_0^2/c^2}}{(1-\beta_0^2-r^2\omega_0^2/c^2)^{1/2}}\right]\right\}, \tag{2.4.22}$$

for $0 < r < R_p$, and $\omega_{pe}^2(r) = 0$ for $r > R_p$. In Eq. (2.4.22), $\omega_{pe}^2(0) = 4\pi n_e^0(r=0)e^2/m_e$, and $\omega_{pe}(0)$, Ω_e, ω_0, R_p, β_0, and f can be related by evaluating Eq. (2.4.3) at $r = R_p$. This gives

$$0 = \frac{\omega_0^2}{(1-\beta_0^2 - R_p^2\omega_0^2/c^2)^{1/2}} + \frac{1}{R_p^2}(1-f-\beta_0^2)\int_0^{R_p} dr'\, r'\omega_{pe}^2(r') - \omega_0\Omega_e, \tag{2.4.23}$$

where $\omega_{pe}^2(r)$ is specified by Eq. (2.4.22). For a fully nonneutral ($f = 0$), nondrifting ($\beta_0^2 = 0$) electron beam, Eq. (2.4.22) reduces to the expression for the electron density profile given in Eq. (2.3.16), as expected.

---

†In particular, Eq. (2.3.7) is identical to Eq. (2.4.3) if 1 is replaced by $1-f-\beta_0^2$ in the coefficient of $r^{-2}\int_0^r dr'\, r'\omega_{pe}^2(r')$.

2.5 MACROSCOPIC MODEL OF THE BENNETT PINCH

In this section the radial force equation (2.1.12) is used to construct a simple macroscopic model of the Bennett pinch.[44] The ions are taken to form a stationary, partially neutralizing background with density

$$n_i^0(r) = f n_e^0(r), \tag{2.5.1}$$

where f = const. = fractional neutralization. It is assumed that the electrons have no mean motion in the azimuthal direction, that is,

$$\beta_{e\theta}(r) = V_{e\theta}^0(r)\,/\,c = 0 \tag{2.5.2}$$

As in Section 2.4, the axial electron velocity is assumed to be independent of r,

$$\beta_{ez}(r) = \beta_0 = \text{const.} \tag{2.5.3}$$

Making use of Eqs. (2.5.1)-(2.5.3), and generalizing the equilibrium force equation (2.1.12) to include a force term due to radial variation in electron pressure $P_e^0(r)$ (assumed to be a scalar), we find

$$\frac{1}{n_e^0(r)}\frac{\partial}{\partial r}P_e^0(r) = \frac{4\pi e^2}{r}(1 - f - \beta_0^2)\int_0^r dr'\, r' n_e^0(r'). \tag{2.5.4}$$

For an isothermal equation of state, $P_e^0(r) = n_e^0(r) k_B T_e$ (T_e = const.), Eq. (2.5.4) can be expressed as

$$\frac{k_B T_e}{m_e}\frac{1}{\omega_{pe}^2(r)}\frac{\partial}{\partial r}\omega_{pe}^2(r) = \frac{1 - f - \beta_0^2}{r}\int_0^r dr'\, r' \omega_{pe}^2(r'), \tag{2.5.5}$$

where $\omega_{pe}^2(r) = 4\pi n_e^0(r) e^2/m_e$. From Eq. (2.5.5) it is seen that for monotonically decreasing density profiles with $\partial\omega_{pe}^2(r)/\partial r \leqslant 0$, the condition

$$\beta_0^2 > 1 - f \tag{2.5.6}$$

is required for radial confinement. The inequality in Eq. (2.5.6) assures that the magnetic pinching forces are larger than the electrostatic repulsive forces. The solution to Eq. (2.5.5) is

$$\omega_{pe}^2(r) = \frac{\omega_{pe}^2(0)}{(1 + r^2/a^2)^2}, \tag{2.5.7}$$

where $\omega_{pe}^2(0) = 4\pi n_e^0(r=0)e^2/m_e$, and

$$a^2 = \frac{8\lambda_D^2}{\beta_0^2 - (1-f)}. \tag{2.5.8}$$

In Eq. (2.5.8) $\lambda_D = [k_B T_e/4\pi n_e^0(r=0)e^2]^{1/2}$ is the electron Debye length at $r=0$. Note that, if β_0^2 is close to $1-f$, the electron beam is many Debye lengths in diameter, that is, $2a \gg \lambda_D$. It is straightforward to show from Eq. (2.5.7) that the density on axis, $n_e^0(r=0)$, is related to a and the number of electrons per unit length of the beam, $\overline{N}_e = 2\pi \int_0^\infty dr'\, r' n_e^0(r')$, by

$$n_e^0(r=0) = \frac{\overline{N}_e}{\pi a^2}. \tag{2.5.9}$$

The bell-shaped density profile given by Eq. (2.5.7) is illustrated in Fig. 2.5.1.

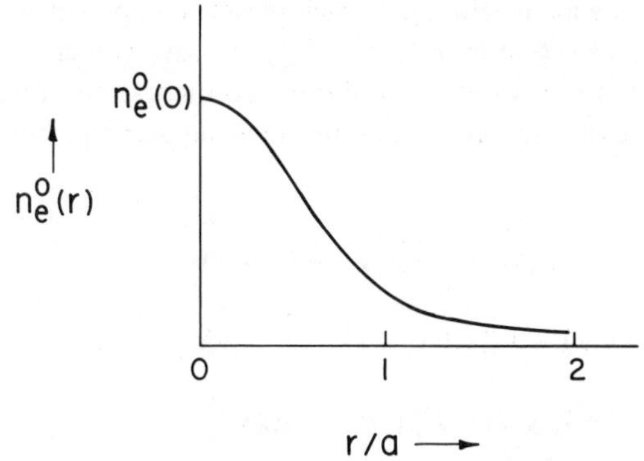

Fig. 2.5.1 Plot of $n_e^0(r)$ versus r/a for the bell-shaped density profile in Eq. (2.5.7).

2.6 ELECTROSTATIC STABILITY OF NONNEUTRAL PLASMAS

In this section the stability of nonrelativistic, nondiamagnetic, nonneutral plasma equilibria (see Section 2.2) to small-amplitude electrostatic perturbations is considered. As in Section 2.1, the plasma column is aligned parallel to a uniform external magnetic field, $\mathbf{B}_0^{\text{ext}}(\mathbf{x}) = B_0\hat{\mathbf{e}}_z$, and it is assumed that the equilibrium is infinite and uniform in the z-direction with $\partial n_\alpha^0(\mathbf{x})/\partial z = 0$, and $\mathbf{E}^0(\mathbf{x}) \cdot \hat{\mathbf{e}}_z = 0$ (see Fig. 2.1.1). The azimuthal and axial motions of the plasma components are assumed to be nonrelativistic with

$$\beta_{\alpha\theta}^2 = r^2 \omega_\alpha^2 / c^2 \ll 1, \qquad (2.6.1)$$

and

$$\beta_{\alpha z}^2 = V_{\alpha z}^{02} / c^2 \ll 1, \qquad (2.6.2)$$

across the radial extent of the nonneutral plasma column. Consistent with Eqs. (2.6.1) and (2.6.2), the equilibrium self magnetic fields, $B_z^s(r)$ and $B_\theta^s(r)$, are omitted from the analysis, and the radial force equation (2.1.12) is approximated by [see Eq. (2.2.6)]

$$\omega_\alpha^2(r) + \sum_\eta \frac{4\pi e_\alpha e_\eta}{m_\alpha r^2} \int_0^r dr'\, r' n_\eta^0(r') + \epsilon_\alpha \Omega_\alpha \omega_\alpha(r) = 0. \qquad (2.6.3)$$

In Eq. (2.6.3), $\omega_\alpha(r) = V_{\alpha\theta}^0(r)/r$ is the angular velocity of mean rotation for component α, r is the radial distance from the axis of rotation, $\epsilon_\alpha = \text{sgn}\, e_\alpha$, $n_\eta^0(r)$ is the radial density profile, and $\Omega_\alpha = |e_\alpha| B_0/m_\alpha c$. In the stability analysis that follows, the axial velocity of each plasma component is taken to be independent of radius r, that is, $V_{\alpha z}^0(r) = V_{\alpha z}^0 =$ const. (independent of r).

To examine the stability of various equilibrium configurations, each quantity of physical interest is expressed as its equilibrium value plus a perturbation, that is,

$$n_\alpha(\mathbf{x}, t) = n_\alpha^0(r) + \delta n_\alpha(\mathbf{x}, t), \qquad (2.6.4a)$$

$$\mathbf{V}_\alpha(\mathbf{x}, t) = V_{\alpha\theta}^0(r)\hat{\mathbf{e}}_\theta + V_{\alpha z}^0\hat{\mathbf{e}}_z + \delta\mathbf{V}_\alpha(\mathbf{x}, t), \qquad (2.6.4b)$$

$$\mathbf{E}(\mathbf{x}, t) = E_r^0(r)\hat{\mathbf{e}}_r + \delta\mathbf{E}(\mathbf{x}, t), \qquad (2.6.4c)$$

$$\mathbf{B}(\mathbf{x}, t) = B_0\hat{\mathbf{e}}_z + \delta\mathbf{B}(\mathbf{x}, t), \qquad (2.6.4d)$$

where $\hat{\mathbf{e}}_r$, $\hat{\mathbf{e}}_\theta$, and $\hat{\mathbf{e}}_z$ are unit vectors in the r-, θ-, and z- directions, respectively (see Fig. 2.1.1). For small-amplitude perturbations, the evolution of $\delta n_\alpha(\mathbf{x}, t)$, $\delta\mathbf{V}_\alpha(\mathbf{x}, t)$, $\delta\mathbf{E}(\mathbf{x}, t)$, and $\delta\mathbf{B}(\mathbf{x}, t)$ is determined from Eqs. (1.3.36)–(1.3.41). In the electrostatic approximation the perturbed magnetic field $\delta\mathbf{B}(\mathbf{x}, t)$ is assumed to remain negligibly small in Eqs. (1.3.37) and (1.3.38), and Eq. (1.3.38) is approximated by

$$\nabla \times \delta\mathbf{E}(\mathbf{x}, t) = 0. \qquad (2.6.5)$$

Equation (2.6.5) implies that the perturbed electric field $\delta\mathbf{E}(\mathbf{x}, t)$ can be expressed as the gradient of a scalar potential,

$$\delta\mathbf{E}(\mathbf{x}, t) = -\nabla\delta\phi(\mathbf{x}, t). \tag{2.6.6}$$

Since the motions of the plasma components are assumed to be nonrelativistic, the perturbed momentum of component α can be approximated in Eq. (1.3.37) by

$$\delta\mathbf{P}_\alpha(\mathbf{x}, t) = m_\alpha\,\delta\mathbf{V}_\alpha(\mathbf{x}, t). \tag{2.6.7}$$

When Eqs. (1.3.36), (1.3.37), (1.3.40), (2.6.4), (2.6.6), and (2.6.7) are combined, the macroscopic fluid-Poisson equations for the perturbed quantities become

$$0 = \frac{\partial}{\partial t}\,\delta n_\alpha(\mathbf{x}, t) + \nabla\cdot\left\{n_\alpha^0(r)\,\delta\mathbf{V}_\alpha(\mathbf{x}, t) + \delta n_\alpha(\mathbf{x}, t)\,[V_{\alpha\theta}^0(r)\hat{\mathbf{e}}_\theta + V_{\alpha z}^0\hat{\mathbf{e}}_z]\right\}, \tag{2.6.8}$$

$$\begin{aligned}\frac{\partial}{\partial t}\delta\mathbf{V}_\alpha(\mathbf{x}, t) &+ [V_{\alpha\theta}^0(r)\hat{\mathbf{e}}_\theta + V_{\alpha z}^0\hat{\mathbf{e}}_z]\cdot\nabla\delta\mathbf{V}_\alpha(\mathbf{x}, t)\\ &+ \delta\mathbf{V}_\alpha(\mathbf{x}, t)\cdot\nabla[V_{\alpha\theta}^0(r)\hat{\mathbf{e}}_\theta]\\ &= \frac{e_\alpha}{m_\alpha}\left[-\nabla\delta\phi(\mathbf{x}, t) + \frac{\delta\mathbf{V}_\alpha(\mathbf{x}, t)\times B_0\hat{\mathbf{e}}_z}{c}\right],\end{aligned} \tag{2.6.9}$$

$$\nabla^2\,\delta\phi(\mathbf{x}, t) = -4\pi\sum_\alpha e_\alpha\,\delta n_\alpha(\mathbf{x}, t). \tag{2.6.10}$$

It is convenient to analyze Eqs. (2.6.8)-(2.6.10) in cylindrical polar coordinates (r, θ, z) with

$$\nabla = \hat{\mathbf{e}}_r\frac{\partial}{\partial r} + \hat{\mathbf{e}}_\theta\,\frac{1}{r}\frac{\partial}{\partial\theta} + \hat{\mathbf{e}}_z\frac{\partial}{\partial z} \tag{2.6.11}$$

and

$$\delta\mathbf{V}_\alpha(\mathbf{x}, t) = \delta V_{\alpha r}(\mathbf{x}, t)\hat{\mathbf{e}}_r + \delta\mathbf{V}_{\alpha\theta}(\mathbf{x}, t)\hat{\mathbf{e}}_\theta + \delta V_{\alpha z}(\mathbf{x}, t)\hat{\mathbf{e}}_z. \tag{2.6.12}$$

To determine the wave and stability properties characteristic of perturbations about equilibrium, a normal-mode approach is adopted. It is assumed that the time variation of perturbed quantities is of the form†

†It is assumed that Im $\omega > 0$ in carrying out the normal-mode analysis of Eqs. (2.6.8)–(2.6.10). The dispersion equation that is obtained can then be defined in the region Im $\omega < 0$ by appropriate analytic continuation. It should be noted that this procedure yields the same result as is obtained when Eqs. (2.6.8)–(2.6.10) are Laplace-transformed with respect to t, and initial-value terms are neglected in the analysis.

$$\exp(-i\omega t). \tag{2.6.13}$$

The complex oscillation frequency ω is determined consistently from Eqs. (2.6.8)-(2.6.10). If Im $\omega > 0$, the perturbations grow and the equilibrium configuration is *unstable*. In analyzing Eqs. (2.6.8)-(2.6.10), the perturbations are assumed to be spatially periodic in the z-direction with periodicity length L. Making use of Eq. (2.6.13), we can Fourier-decompose the θ- and z-dependences of all perturbed quantities according to

$$\delta\psi(r,\theta,z,t) = \sum_{\ell=-\infty}^{\infty} \sum_{k_z=-\infty}^{\infty} \delta\psi^{\ell}(r,k_z) \exp[i(\ell\theta + k_z z - \omega t)], \tag{2.6.14}$$

where $k_z = 2\pi n/L$, and n is an integer. Substituting Eq. (2.6.14) into Eqs. (2.6.8)-(2.6.10), and making use of Eqs. (2.6.11) and (2.6.12) we can show that the Fourier amplitudes $\delta n_\alpha^\ell(r,k_z)$, $\delta V_{\alpha r}^\ell(r,k_z)$, $\delta V_{\alpha\theta}^\ell(r,k_z)$, $\delta V_{\alpha z}^\ell(r,k_z)$, and $\delta\phi^\ell(r,k_z)$ satisfy†

$$-i(\omega - k_z V_{\alpha z}^0 - \ell\omega_\alpha)\,\delta n_\alpha^\ell + \frac{1}{r}\frac{\partial}{\partial r}(r n_\alpha^0\,\delta V_{\alpha r}^\ell) + \frac{i\ell n_\alpha^0\,\delta V_{\alpha\theta}^\ell}{r} + ik_z n_\alpha^0\,\delta V_{\alpha z}^\ell = 0, \tag{2.6.15}$$

$$-i(\omega - k_z V_{\alpha z}^0 - \ell\omega_\alpha)\,\delta V_{\alpha r}^\ell - (\epsilon_\alpha\Omega_\alpha + 2\omega_\alpha)\,\delta V_{\alpha\theta}^\ell = -\frac{e_\alpha}{m_\alpha}\frac{\partial}{\partial r}\delta\phi^\ell, \tag{2.6.16}$$

$$-i(\omega - k_z V_{\alpha z}^0 - \ell\omega_\alpha)\,\delta V_{\alpha\theta}^\ell + \left[\epsilon_\alpha\Omega_\alpha + \frac{1}{r}\frac{\partial}{\partial r}(r^2\omega_\alpha)\right]\delta V_{\alpha r}^\ell = -\frac{e_\alpha}{m_\alpha}\frac{i\ell\,\delta\phi^\ell}{r}, \tag{2.6.17}$$

$$-i(\omega - k_z V_{\alpha z}^0 - \ell\omega_\alpha)\,\delta V_{\alpha z}^\ell = -\frac{e_\alpha}{m_\alpha} ik_z\,\delta\phi^\ell, \tag{2.6.18}$$

$$\frac{1}{r}\frac{\partial}{\partial r} r \frac{\partial}{\partial r}\delta\phi^\ell - \frac{\ell^2}{r^2}\delta\phi^\ell - k_z^2\,\delta\phi^\ell = -4\pi\sum_\alpha e_\alpha\,\delta n_\alpha^\ell, \tag{2.6.19}$$

where the (r, k_z) arguments of the perturbations have been suppressed. In Eqs. (2.6.15)-(2.6.19), $\epsilon_\alpha = \text{sgn}\, e_\alpha$, $\Omega_\alpha = |e_\alpha| B_0/m_\alpha c$, and the equilibrium angular velocity profile, $\omega_\alpha(r) = V_{\alpha\theta}^0(r)/r$, is related to the equilibrium density profile $n_\alpha^0(r)$ by the radial force equation (2.6.3). The perturbations in density and mean fluid velocities in Eqs. (2.6.15)-(2.6.18) can be eliminated in favor of $\delta\phi^\ell(r,k_z)$. Poisson's equation for the perturbed electrostatic potential can then be expressed in the form

†For the representation in Eq. (2.6.14), note that $\frac{\partial}{\partial t} \to -i\omega$, $\frac{\partial}{\partial\theta} \to i\ell$, and $\frac{\partial}{\partial z} \to ik_z$ for each Fourier component.

$$\frac{1}{r}\frac{\partial}{\partial r}\left[r\left(1-\sum_{\alpha}\frac{\omega_{p\alpha}^2}{\nu_{\alpha}^2}\right)\frac{\partial}{\partial r}\delta\phi^{\ell}\right]$$

$$-\frac{\ell^2}{r^2}\left(1-\sum_{\alpha}\frac{\omega_{p\alpha}^2}{\nu_{\alpha}^2}\right)\delta\phi^{\ell}-k_z^2\left[1-\sum_{\alpha}\frac{\omega_{p\alpha}^2}{(\omega-k_zV_{\alpha z}^0-\ell\omega_{\alpha})^2}\right]\delta\phi^{\ell}$$

$$=-\frac{\ell\,\delta\phi^{\ell}}{r}\sum_{\alpha}\frac{1}{\omega-k_zV_{\alpha z}^0-\ell\omega_{\alpha}}\frac{\partial}{\partial r}\left[\frac{\omega_{p\alpha}^2}{\nu_{\alpha}^2}(\epsilon_{\alpha}\Omega_{\alpha}+2\omega_{\alpha})\right], \tag{2.6.20}$$

where

$$\nu_{\alpha}^2(r)\equiv(\omega-k_zV_{\alpha z}^0-\ell\omega_{\alpha})^2-(\epsilon_{\alpha}\Omega_{\alpha}+2\omega_{\alpha})\left[\epsilon_{\alpha}\Omega_{\alpha}+\frac{1}{r}\frac{\partial}{\partial r}(r^2\omega_{\alpha})\right], \tag{2.6.21}$$

and $\omega_{p\alpha}^2(r)\equiv 4\pi n_{\alpha}^0(r)e_{\alpha}^2/m_{\alpha}$. Equation (2.6.20) is valid for arbitrary $\omega_{p\alpha}^2(r)$ and $\omega_{\alpha}(r)$ consistent with the radial force equation (2.6.3). Operationally the procedure is to solve Eq. (2.6.20) for $\delta\phi^{\ell}(r, k_z)$ and ω as an eigenvalue problem. The solution to Eq. (2.6.20) is accessible analytically only for certain simple density profiles. As a first application of Eq. (2.6.20), the case in which the density of each plasma component is constant in the column interior is considered.

2.7 DISPERSION RELATION FOR A CONSTANT-DENSITY PLASMA COLUMN

In this section use is made of Eq. (2.6.20) to obtain the dispersion relation for electrostatic waves in a constant-density nonneutral plasma column.[3–8, 140] As illustrated in Figure 2.7.1, it is assumed that the density of each plasma component is constant out to a radius $r=R_p$ and is zero beyond, that is,

$$n_{\alpha}^0(r)=\begin{cases}\bar{n}_{\alpha}, & 0<r<R_p,\\ 0, & R_p<r<R_c.\end{cases} \tag{2.7.1}$$

In Eq. (2.7.1), $r=R_c\geqslant R_p$ denotes the radial location of a perfectly conducting wall (see Fig. 2.7.1). It follows from Eq. (2.7.1) that $\omega_{p\alpha}^2(r)=4\pi n_{\alpha}^0(r)e_{\alpha}^2/m_{\alpha}$ is constant for each plasma component, with

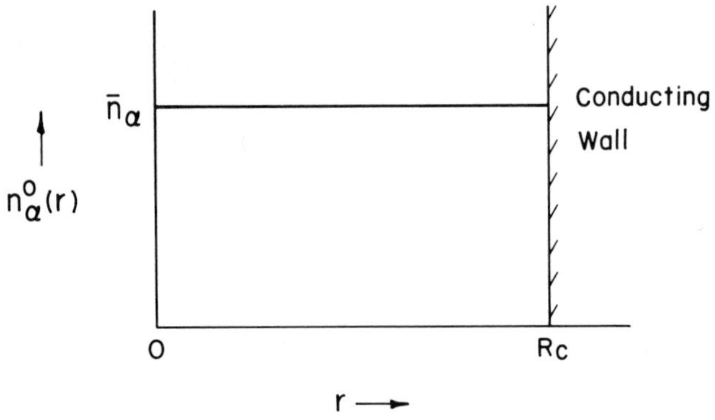

Fig. 2.7.1 Plot of $n^0_\alpha(r)$ versus r for the constant-density profile in Eq. (2.7.1). A grounded conducting wall is located at radius $r = R_c \geqslant R_p$.

$$\omega^2_{p\alpha}(r) = \begin{cases} \dfrac{4\pi\bar{n}_\alpha e^2_\alpha}{m_\alpha} \equiv \bar{\omega}^2_{p\alpha}, & 0 < r < R_p, \\ 0, & R_p < r < R_c. \end{cases} \tag{2.7.2}$$

Furthermore, it follows from Eqs. (2.6.3) and (2.7.1) that the equilibrium rotation velocity $\omega_\alpha(r)$ in the region $0 < r < R_p$ can be expressed as [see Eq. (2.2.12)]

$$\omega_\alpha(r) = \omega^\pm_\alpha \equiv -\frac{\epsilon_\alpha \Omega_\alpha}{2}\left\{1 \pm \left[1 - 2\sum_\eta \frac{4\pi e_\alpha e_\eta \bar{n}_\eta}{m_\alpha \Omega^2_\alpha}\right]^{1/2}\right\} \tag{2.7.3}$$

in the nonrelativistic, nondiamagnetic regime. Note from Eq. (2.7.3) that ω_α is independent of radius r. Therefore it follows from Eq. (2.6.21) that ν^2_α can be expressed as

$$\nu^2_\alpha = (\omega - k_z V^0_{\alpha z} - \ell\omega_\alpha)^2 - (\epsilon_\alpha \Omega_\alpha + 2\omega_\alpha)^2, \tag{2.7.4}$$

which is also independent of r. For the density profile given in Eq. (2.7.2),

$$\frac{\partial}{\partial r}\omega^2_{p\alpha} = -\bar{\omega}^2_{p\alpha}\,\delta(r - R_p). \tag{2.7.5}$$

Substituting Eq. (2.7.5) into Eq. (2.6.20), and making use of the constancy of ν_α^2 and ω_α, we find that Poisson's equation for the perturbed potential $\delta\phi^\ell(r, k_z)$, can be expressed as

$$\frac{1}{r}\frac{\partial}{\partial r}\left[r\left(1-\sum_\alpha \frac{\omega_{p\alpha}^2}{\nu_\alpha^2}\right)\frac{\partial}{\partial r}\delta\phi^\ell\right]$$

$$-\frac{\ell^2}{r^2}\left(1-\sum_\alpha \frac{\omega_{p\alpha}^2}{\nu_\alpha^2}\right)\delta\phi^\ell - k_z^2\left[1-\sum_\alpha \frac{\omega_{p\alpha}^2}{(\omega-k_zV_{\alpha z}^0-\ell\omega_\alpha)^2}\right]\delta\phi^\ell$$

$$=\ell\,\frac{\delta\phi^\ell}{r}\sum_\alpha \frac{\bar{\omega}_{p\alpha}^2(\epsilon_\alpha\Omega_\alpha+2\omega_\alpha)}{\nu_\alpha^2(\omega-k_zV_{\alpha z}^0-\ell\omega_\alpha)}\,\delta(r-R_p), \tag{2.7.6}$$

where ω_α and ν_α^2 are defined in Eqs. (2.7.3) and (2.7.4). The right-hand side of Eq. (2.7.6), which is zero except at $r = R_p$, corresponds to a perturbation in charge density on the surface of the plasma column.

Inside the plasma column ($0 < r < R_p$), where $\omega_{p\alpha}^2(r) = \bar{\omega}_{p\alpha}^2 = \text{const.}$ [see Eq. (2.7.2)], Eq. (2.7.6) can be expressed in the form

$$\frac{1}{r}\frac{\partial}{\partial r}r\frac{\partial}{\partial r}\delta\phi^\ell - \frac{\ell^2}{r^2}\delta\phi^\ell + T^2\,\delta\phi^\ell = 0, \quad 0<r<R_p, \tag{2.7.7}$$

where

$$T^2 \equiv -k_z^2\,\frac{\left[1-\sum_\alpha \dfrac{\bar{\omega}_{p\alpha}^2}{(\omega-k_zV_{z\alpha}^0-\ell\omega_\alpha)^2}\right]}{\left[1-\sum_\alpha \dfrac{\bar{\omega}_{p\alpha}^2}{\nu_\alpha^2}\right]}. \tag{2.7.8}$$

Outside the plasma column ($R_p < r < R_c$), where $\omega_{p\alpha}^2(r) = 0$ [see Eq. (2.7.2)], Eq. (2.7.6) reduces to Poisson's equation in free space, that is,

$$\frac{1}{r}\frac{\partial}{\partial r}r\frac{\partial}{\partial r}\delta\phi^\ell - \frac{\ell^2}{r^2}\delta\phi^\ell - k_z^2\,\delta\phi^\ell = 0, \quad R_p<r<R_c. \tag{2.7.9}$$

Equations (2.7.7) and (2.7.9) are both forms of Bessel's equation. The solution to Eq. (2.7.7) that remains finite at $r = 0$ is

$$\delta\phi_{\text{in}}^\ell = AJ_\ell(Tr), \tag{2.7.10}$$

where J_ℓ is the Bessel function of the first kind of order ℓ, and A = const. (independent of r). The solution to Eq. (2.7.9) that vanishes at the conducting wall ($r = R_c$) is

$$\delta\phi^\ell_{\text{out}} = C[I_\ell(k_z r)K_\ell(k_z R_c) - K_\ell(k_z r)I_\ell(k_z R_c)]\,, \quad R_p < r < R_c, \tag{2.7.11}$$

where I_ℓ and K_ℓ are modified Bessel functions of order ℓ, and C = const. (independent of r).

Equations (2.7.10) and (2.7.11) constitute the solutions for the perturbed electrostatic potentials $\delta\phi^\ell$ in the inner and outer regions, respectively. The boundary conditions at the surface of the plasma column ($r = R_p$) remain to be enforced. Continuity of $\delta\phi^\ell$ at $r = R_p$ gives†

$$[\delta\phi^\ell_{\text{in}}]_{r=R_p} = [\delta\phi^\ell_{\text{out}}]_{r=R_p}, \tag{2.7.12}$$

which connects the solutions in the two regions. Substituting Eqs. (2.7.10) and (2.7.11) into Eq. (2.7.12) gives

$$\frac{C}{A} = \frac{J_\ell(TR_p)}{I_\ell(k_z R_p)K_\ell(k_z R_c) - K_\ell(k_z R_p)I_\ell(k_z R_c)}, \tag{2.7.13}$$

which relates the constants C and A. A further relation between $\delta\phi^\ell_{\text{in}}$ and $\delta\phi^\ell_{\text{out}}$ is obtained by multiplying Eq. (2.7.6) by r, integrating from $r = R_p - \epsilon$ to $r = R_p + \epsilon$, and taking the limit $\epsilon \to 0_+$. Making use of Eq. (2.7.12), we obtain

$$R_p\left[\frac{\partial}{\partial r}\delta\phi^\ell_{\text{out}}\right]_{r=R_p} - R_p\left[1 - \sum_\alpha \frac{\overline{\omega}^2_{p\alpha}}{\nu^2_\alpha}\right]\left[\frac{\partial}{\partial r}\delta\phi^\ell_{\text{in}}\right]_{r=R_p}$$
$$= \ell\,[\delta\phi^\ell]_{r=R_p} \sum_\alpha \frac{\overline{\omega}^2_{p\alpha}(\epsilon_\alpha\Omega_\alpha + 2\omega_\alpha)}{\nu^2_\alpha(\omega - k_z V^0_{\alpha z} - \ell\omega_\alpha)}, \tag{2.7.14}$$

Equation (2.7.14) relates the discontinuity in perturbed radial electric field at $r = R_p$ to the surface charge density produced by the perturbation. Substituting Eqs. (2.7.10), (2.7.11), and (2.7.13) into Eq. (2.7.14) gives

---

†Since the Fourier amplitude for the perturbed azimuthal electric field can be expressed as $\delta E^0_\theta = -i\ell\,\delta\phi^\ell/r$, Eq. (2.7.12) is equivalent to the continuity of tangential electric field at $r = R_p$.

$$k_z R_p \frac{K_\ell(k_z R_c) I'_\ell(k_z R_p) - K'_\ell(k_z R_p) I_\ell(k_z R_c)}{K_\ell(k_z R_c) I_\ell(k_z R_p) - K_\ell(k_z R_p) I_\ell(k_z R_c)}$$

$$- \left(1 - \sum_\alpha \frac{\overline{\omega}_{p\alpha}^2}{\nu_\alpha^2}\right) TR_p \frac{J'_\ell(TR_p)}{J_\ell(TR_p)} = \ell \sum_\alpha \frac{\overline{\omega}_{p\alpha}^2 (\epsilon_\alpha \Omega_\alpha + 2\omega_\alpha)}{\nu_\alpha^2 (\omega - k_z V_{\alpha z}^0 - \ell\omega_\alpha)} . \tag{2.7.15}$$

The "prime" notation in Eq. (2.7.15) denotes derivatives with respect to the *complete* argument of the Bessel function, for example,

$$I'_\ell(k_z R_p) = [dI_\ell(x)/dx]_{x = k_z R_p}.$$

Equation (2.7.15) is the *dispersion relation* for electrostatic waves in a cold constant-density, nonneutral plasma column. It relates the (complex) oscillation frequency ω to the azimuthal harmonic number ℓ, the wave vector k_z and properties characteristic of the equilibrium configuration (e.g., $\overline{\omega}_{p\alpha}^2$, ω_α R_p, and R_c). In general, it is not possible to obtain a closed analytic expression for ω from Eq. (2.7.15). However, Eq. (2.7.15) does simplify in various limiting cases, which include the following:

(*a*) $R_p = R_c$ (plasma-filled waveguide),
(*b*) $k_z^2 R_p^2 \ll 1$ (limit of long axial wavelengths),
(*c*) $R_p \ll R_c$ (limit of a thin beam).

In Sections 2.8, 2.9, and 2.11, Eq. (2.7.15) is analyzed in these limiting cases to determine the dispersive properties of electrostatic waves propagating in a constant-density plasma column.

2.8 NONNEUTRAL PLASMA-FILLED WAVEGUIDE

2.8.1 Electrostatic Dispersion Relation

In this section the dispersive properties of electrostatic waves propagating in a nonneutral plasma-filled waveguide are discussed. As illustrated in Fig. 2.8.1, it is assumed that the density of each plasma component is constant, and that the radius of the plasma column extends to the conducting wall, that is,

$$R_p = R_c. \tag{2.8.1}$$

The electrostatic dispersion relation for this equilibrium configuration can be obtained by (*a*) taking the limit $R_p \to R_c$ in the dispersion relation, Eq. (2.7.15),

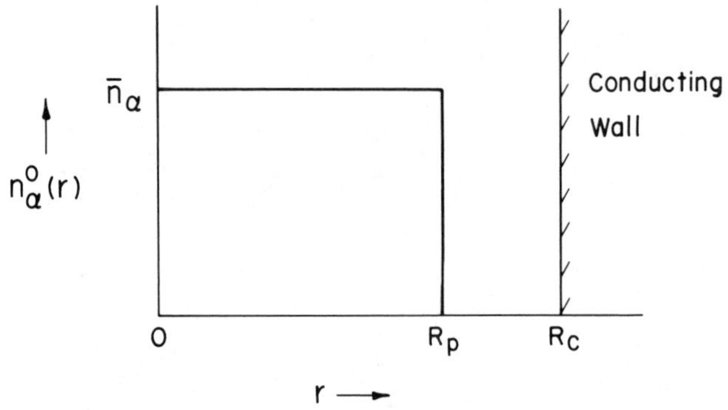

Fig. 2.8.1 Plot of $n_\alpha^0(r)$ versus r for a constant-density plasma-filled waveguide $(R_p = R_c)$.

or (*b*) requiring that the perturbed electrostatic potential vanish at the conducting wall, that is, $[\delta\phi^\ell]_{r=R_c} = 0$ in Eq. (2.7.10). Either approach leads to the condition

$$J_\ell(TR_c) = 0, \tag{2.8.2}$$

where T is defined in Eq. (2.7.8). It follows from Eq. (2.8.2) that

$$T^2R_c^2 = p_{\ell m}^2, \quad m = 1, 2, \cdots, \tag{2.8.3}$$

where $p_{\ell m}$ is the mth zero of $J_\ell(x) = 0$. Making use of Eq. (2.7.8), and introducing the effective perpendicular wavenumber $k_\perp$, where

$$k_\perp^2 \equiv \frac{p_{\ell m}^2}{R_c^2}, \tag{2.8.4}$$

we can express Eq. (2.8.3) in the equivalent form

$$0 = 1 - \frac{k_\perp^2}{k^2}\sum_\alpha \frac{\bar{\omega}_{p\alpha}^2}{(\omega - \ell\omega_\alpha - k_zV_{\alpha z}^0)^2 - \omega_{\alpha v}^2} - \frac{k_z^2}{k^2}\sum_\alpha \frac{\bar{\omega}_{p\alpha}^2}{(\omega - \ell\omega_\alpha - k_zV_{\alpha z}^0)^2}. \tag{2.8.5}$$

In Eq. (2.8.5), $\bar{\omega}_{p\alpha}^2 = 4\pi\bar{n}_\alpha e_\alpha^2/m_\alpha$, $k^2 \equiv k_z^2 + k_\perp^2$, ω_α is the angular velocity of mean rotation defined in Eq. (2.7.3), and $\omega_{\alpha v}$ is the *vortex frequency*,[8] defined by

$$\omega_{\alpha v} \equiv -(\epsilon_\alpha \Omega_\alpha + 2\omega_\alpha). \tag{2.8.6}$$

Making use of Eqs. (2.7.3) and (2.8.6), we can express $\omega_{\alpha v}^2$ as

$$\omega_{\alpha v}^2 \equiv (\omega_\alpha^+ - \omega_\alpha^-)^2. \tag{2.8.7}$$

Equation (2.8.5) is the dispersion relation for electrostatic waves in a nonneutral plasma-filled waveguide. Note that the allowed values of $k_\perp$ in Eq. (2.8.5) are quantized by the effects of finite radial geometry [Eq. (2.8.4)]. The physical significance of the vortex frequency $\omega_{\alpha v}$ which appears in Eq. (2.8.5) can be summarized as follows. If a particle of species α is perturbed about an axicentered circular orbit, then in a frame of reference rotating with angular velocity ω_α the *perturbed orbits* are circular gyrations with angular frequency $\omega_{\alpha v}$ (see Section 1.2). It is important to note that Eq. (2.8.5) is similar in form to the dispersion relation for a neutral plasma-filled waveguide. In particular, Eq. (2.8.5) is reproduced identically if the replacements,

$$\omega \to \omega - \ell\omega_\alpha \tag{2.8.8}$$

and

$$\Omega_\alpha^2 \to \omega_{\alpha v}^2, \tag{2.8.9}$$

are made in the corresponding dispersion relation for a neutral plasma. Physically, it is very plausible that Eqs. (2.8.8) and (2.8.9) are the correct algorithms for recovering the nonneutral plasma dispersion relation from the neutral plasma dispersion relation. In the nonneutral case, component α is rotating with angular velocity ω_α; hence ω is Doppler-shifted to $\omega - \ell\omega_\alpha$ for spatial perturbations with azimuthal harmonic number ℓ. Moreover, in a frame rotating with angular velocity ω_α, the particles in a nonneutral plasma are gyrating with angular frequency $\omega_{\alpha v}$; hence Ω_α^2 is replaced by $\omega_{\alpha v}^2$. Of course, the dispersion relation for a neutral plasma-filled waveguide can be obtained directly from Eq. (2.8.5) in the limit of equilibrium charge neutrality. Assuming $\omega_\alpha = \omega_\alpha^-$, it is readily verified from Eqs. (2.7.3) and (2.8.6) that $\omega_\alpha = 0$ and $\omega_{\alpha v}^2 = \Omega_\alpha^2$ when $\Sigma_\eta\, e_\eta \bar{n}_\eta = 0$. Hence, for a neutral plasma, Eq. (2.8.5) reduces to

$$0 = 1 - \frac{k_\perp^2}{k^2} \sum_\alpha \frac{\bar{\omega}_{p\alpha}^2}{(\omega - k_z V_{\alpha z}^0)^2 - \Omega_\alpha^2} - \frac{k_z^2}{k^2} \sum_\alpha \frac{\bar{\omega}_{p\alpha}^2}{(\omega - k_z V_{\alpha z}^0)^2}. \tag{2.8.10}$$

Equation (2.8.10) is the correct dispersion relation for a neutral plasma-filled waveguide[97] in which the plasma components have no mean azimuthal motion.†

The neutral plasma dispersion relation, Eq. (2.8.10), has been extensively studied in the literature. Because of the similarity in structure between Eq. (2.8.10) and the nonneutral plasma dispersion relation, Eq. (2.8.5), the waves and instabilities characteristic of a neutral plasma-filled waveguide have their analogs in the nonneutral case. Moreover, many of the quantitative results for a neutral plasma (e.g., dispersion curves and stability criteria) can be applied virtually intact in the nonneutral case, making use of the algorithms given in Eqs. (2.8.8) and (2.8.9). Because of the similarity between Eqs. (2.8.5) and (2.8.10), no attempt will be made here to catalog all of the interesting waves and instabilities characteristic of nonneutral plasma-filled waveguides. Rather, it is sufficient for present purposes to consider two examples that illustrate (*a*) stable oscillations with $\mathrm{Im}\,\omega = 0$, and (*b*) instability with $\mathrm{Im}\,\omega > 0$, in a nonneutral plasma-filled waveguide.

2.8.2 *Stable Oscillations*

As an application of the dispersion relation, Eq. (2.8.5), which corresponds to stable oscillations with $\mathrm{Im}\,\omega = 0$, consider the limit of a fixed ($m_i \to \infty$), partially neutralizing, ion background‡ with density

$$\bar{n}_i = f\bar{n}_e. \tag{2.8.11}$$

For a *single* component of electrons, Eq. (2.8.5) reduces to

$$0 = 1 - \frac{k_\perp^2}{k^2}\frac{\overline{\omega}_{pe}^2}{(\omega - \ell\omega_e - k_z V_{ez}^0)^2 - \omega_{ev}^2} - \frac{k_z^2}{k^2}\frac{\overline{\omega}_{pe}^2}{(\omega - \ell\omega_e - k_z V_{ez}^0)^2}, \tag{2.8.12}$$

where

$$\omega_e = \omega_e^{\pm} = \frac{\Omega_e}{2}\left\{1 \pm \left[1 - \frac{2\overline{\omega}_{pe}^2}{\Omega_e^2}(1 - f)\right]^{1/2}\right\}, \tag{2.8.13}$$

---

†If any of the plasma components is rotating in the *fast* rotational mode ($\omega_\alpha = \omega_\alpha^+$), then $\omega_\alpha = -\epsilon_\alpha \Omega_\alpha$ in the limit where $\Sigma_\eta e_\eta \bar{n}_\eta = 0$ [see Eq. (2.7.3)]. In this case ω must be Doppler-shifted to $\omega + \ell\epsilon_\alpha\Omega_\alpha$ for the fast components in Eq. (2.8.10).

‡Physically, a *fixed* ion background corresponds to examining Eq. (2.8.5) for frequencies ω far removed from any ion resonances.

$$\omega_{ev}^2 = (\omega_e^+ - \omega_e^-)^2 = \Omega_e^2 \left[1 - \frac{2\bar{\omega}_{pe}^2}{\Omega_e^2}(1-f)\right]. \qquad (2.8.14)$$

The solution to Eq. (2.8.12) can be expressed as

$$(\omega - \ell\omega_e - k_z V_{ez}^0)^2$$
$$= \frac{\bar{\omega}_{pe}^2 + \omega_{ev}^2}{2} \left\{1 \pm \left[1 - \frac{k_z^2/k_\perp^2}{1 + k_z^2/k_\perp^2} \frac{4\bar{\omega}_{pe}^2\omega_{ev}^2}{(\bar{\omega}_{pe}^2 + \omega_{ev}^2)^2}\right]^{1/2}\right\}. \qquad (2.8.15)$$

If the equilibrium corresponds to a slow rotational mode ($\omega_e = \omega_e^-$), and charge neutralization is complete ($f = 1$), Eq. (2.8.15) reduces to the correct result for a netural plasma-filled waveguide,[97] that is,

$$(\omega - k_z V_{ez}^0)^2 = \frac{\bar{\omega}_{pe}^2 + \Omega_e^2}{2} \left\{1 \pm \left[1 - \frac{k_z^2/k_\perp^2}{1 + k_z^2/k_\perp^2} \frac{4\bar{\omega}_{pe}^2\Omega_e^2}{(\bar{\omega}_{pe}^2 + \Omega_e^2)^2}\right]^{1/2}\right\}, \qquad (2.8.16)$$

as expected.

In the dispersion relation, Eq. (2.8.15), it is convenient to define

$$\omega' = \omega - \ell\omega_e - k_z V_{ez}^0, \qquad (2.8.17)$$

where ω' is the frequency of the perturbation viewed in a frame comoving with the electron equilibrium configuration. The dispersion curves associated with the normal modes in Eq. (2.8.15) are shown in Figs. 2.8.2 and 2.8.3, where ω'^2 is ploted versus $k_z/k_\perp$. The two cases, $\omega_{ev}^2 > \bar{\omega}_{pe}^2$ (Fig. 2.8.2) and $\omega_{ev}^2 < \bar{\omega}_{pe}^2$ (Fig. 2.8.3), are distinguished. From Eq. (2.8.14) it is straightforward to show that

$$\omega_{ev}^2 > \bar{\omega}_{pe}^2 \text{ is equivalent to } \frac{2\bar{\omega}_{pe}^2}{\Omega_e^2} < \frac{2}{3-2f}, \qquad (2.8.18)$$

and

$$\omega_{ev}^2 < \bar{\omega}_{pe}^2 \text{ is equivalent to } \frac{2\bar{\omega}_{pe}^2}{\Omega_e^2} > \frac{2}{3-2f}. \qquad (2.8.19)$$

Of course, in order for the present analysis to be valid, the value of $2\bar{\omega}_{pe}^2/\Omega_e^2$ must be less than the maximum density for which equilibrium solutions exist. The upper bound on density if obtained by requiring that the radical in Eq. (2.8.13) be real; this gives the condition

$$\frac{2\bar{\omega}_{pe}^2}{\Omega_e^2} \leqslant \frac{1}{1-f}. \qquad (2.8.20)$$

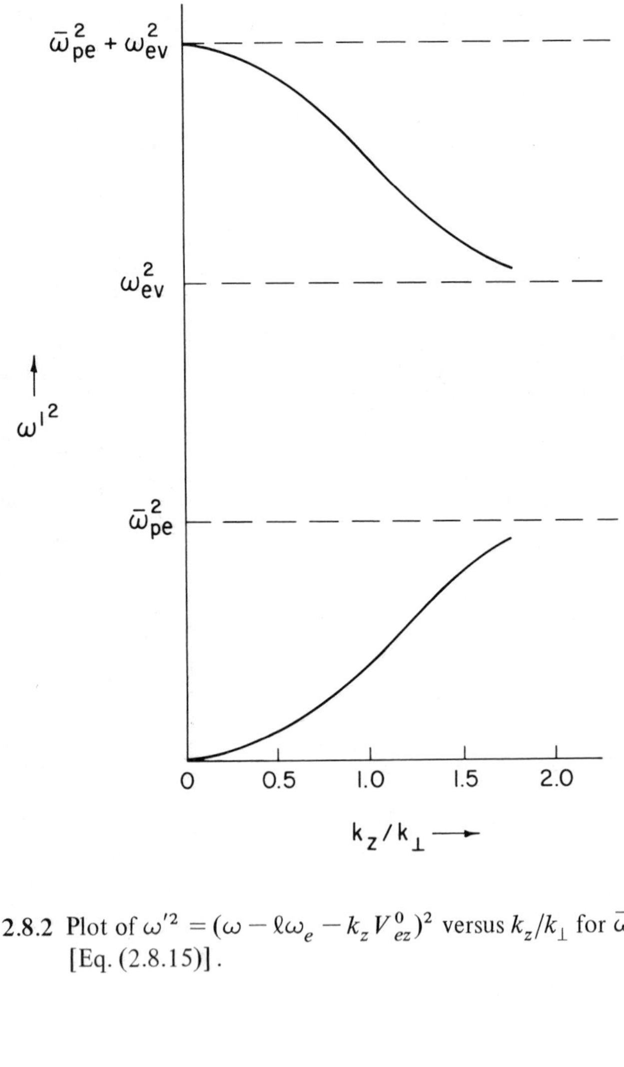

Fig. 2.8.2 Plot of $\omega'^2 = (\omega - \ell\omega_e - k_z V_{ez}^0)^2$ versus $k_z/k_\perp$ for $\bar{\omega}_{pe}^2/\omega_{ev}^2 = 1/2$ [Eq. (2.8.15)].

It should be noted from Eq. (2.8.15) that the only dependence of ω'^2 on $k_\perp^2 = p_{\ell m}^2/R_c^2$ is through the ratio $k_z^2/k_\perp^2$. Therefore the dispersion curves in Figs. 2.8.2 and 2.8.3 are *universal,* that is, for fixed values of $\bar{\omega}_{pe}^2$ and ω_{ev}^2, the curves are identical for all values of $k_\perp^2$.

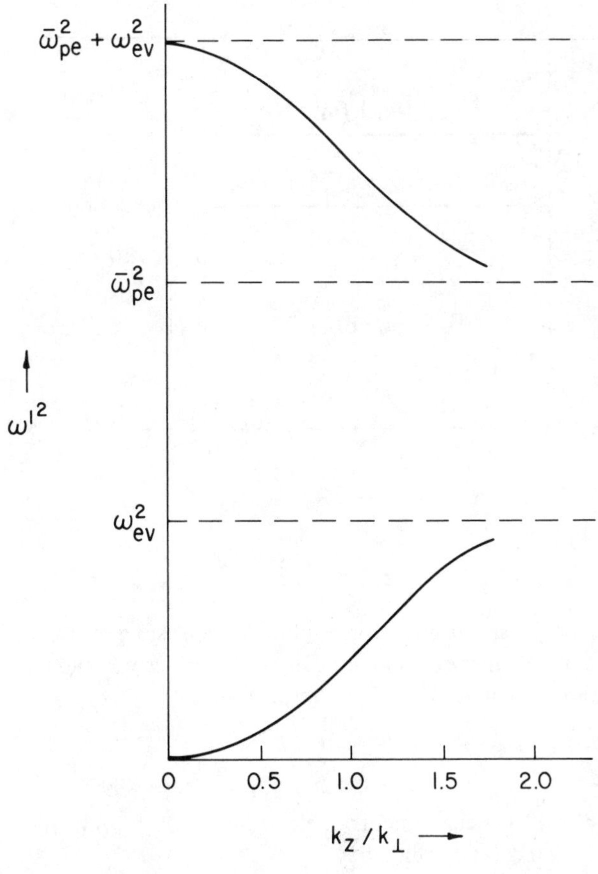

Fig. 2.8.3 Plot of $\omega'^2 = (\omega - \ell\omega_e - k_z V_{ez}^0)^2$ versus $k_z/k_\perp$ for $\bar{\omega}_{pe}^2/\omega_{ev}^2 = 2$ [Eq. (2.8.15)].

2.8.3 Electron-Electron Two-Rotating-Stream Instability

As an application of the dispersion relation, Eq. (2.8.5), which corresponds to instability, consider the equilibrium configuration illustrated in Fig. 2.8.4†. The

† The choice of equilibrium configuration with *two* electron components can be motivated as follows. Suppose that a nonneutral plasma equilibrium is already established with electrons rotating in the *fast* rotational mode, $\omega_e = \omega_e^+$. Suppose also that a low-density background gas of netural atoms is present, as will be the case in a laboratory experiment. As ionization of the background gas proceeds, the electrons and ions produced in this process are essentially "born at rest." Under the influence of the radial electric field it is expected that the electrons produced in the ionization process begin to $\mathbf{E}^0 \times \mathbf{B}_0$ rotate in the slow rotational mode with angular velocity $\omega_e = \omega_e^-$.

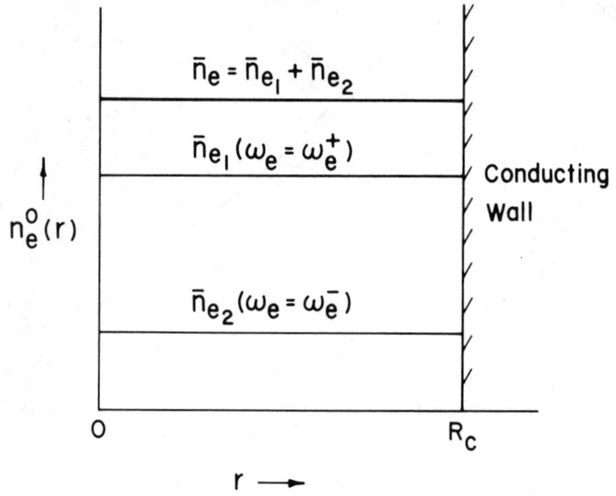

Fig. 2.8.4 Electron density profiles for equilibrium consisting of *two* electron components ($\alpha = e_1, e_2$) with constant densities $\bar{n}_{e_1}$ and $\bar{n}_{e_2}$ and rotation velocities ω_e^+ and ω_e^-, respectively.

equilibrium consists of *two* components of electrons ($\alpha = e_1, e_2$) with constant densities $\bar{n}_{e_1}, \bar{n}_{e_2}$, and rotation velocities

$$\begin{aligned} \omega_\alpha &= \omega_e^+, \quad \text{for} \quad \alpha = e_1, \\ \omega_\alpha &= \omega_e^-, \quad \text{for} \quad \alpha = e_2. \end{aligned} \tag{2.8.21}$$

As in Section 2.8.2, the positive ions are assumed to form a fixed ($m_i \to \infty$), partially neutralizing background with density

$$\bar{n}_i = f\bar{n}_e = f(\bar{n}_{e_1} + \bar{n}_{e_2}). \tag{2.8.22}$$

The angular velocities of the two electron components, ω_e^+ and ω_e^-, are given by Eq. (2.7.3) with $\bar{\omega}_{pe}^2$ defined in terms of the *total* electron density, that is,

$$\overline{\omega}_{pe}^2 = \overline{\omega}_{pe_1}^2 + \overline{\omega}_{pe_2}^2 = \frac{4\pi e^2}{m_e}(\overline{n}_{e_1} + \overline{n}_{e_2}). \tag{2.8.23}$$

For present purposes, it is also assumed that there is no axial drift of the electron components,

$$V_{\alpha z}^0 = 0, \quad \alpha = e_1, e_2. \tag{2.8.24}$$

Because of the differential rotation of the two electron components, it is anticipated that free energy is available to drive an electron-electron two-rotating-stream instability,[79] at least in certain parameter regimes. In the limit $m_i \to \infty$, Eq. (2.8.5) reduces to

$$0 = 1 - \frac{k_\perp^2}{k^2} \sum_{\alpha = e_1, e_2} \frac{\overline{\omega}_{p\alpha}^2}{(\omega - \ell\omega_\alpha)^2 - \omega_{ev}^2}$$

$$- \frac{k_z^2}{k^2} \sum_{\alpha = e_1, e_2} \frac{\overline{\omega}_{p\alpha}^2}{(\omega - \ell\omega_\alpha)^2}, \tag{2.8.25}$$

where ω_{ev}^2 is defined in Eq. (2.8.14), and use has been made of Eq. (2.8.24).

In general, closed solutions for the complex oscillation frequency ω are not accessible from Eq. (2.8.25). However, the condition for Eq. (2.8.25) to support unstable solutions ($\text{Im}\,\omega = \omega_i > 0$) can be found exactly for the case of equi-density electron components,

$$\overline{n}_{e_1} = \overline{n}_{e_2}. \tag{2.8.26}$$

Making use of Eqs. (2.8.25) and (2.8.26), we obtain for the instability condition[79]

$$\frac{4\overline{\omega}_{pe}^2}{(\omega_e^+ - \omega_e^-)^2}\left[\frac{k_z^2}{k^2}\frac{1}{\ell^2} + \frac{k_\perp^2}{k^2}\frac{1}{(\ell^2 - 4 - \ell^4\delta_{|\ell|,2})}\right] > 1 \;\; (\text{for } \ell = \pm 1, \pm 2, \ldots), \tag{2.8.27}$$

where $\overline{\omega}_{pe}^2$ is defined in Eq. (2.8.23), and δ_{ij} is the Kronecker delta.

For $\ell = 0$, the solutions to Eq. (2.8.25) have $\text{Im}\,\omega = 0$, and there is no instability. For $\ell = \pm 1$, and $\overline{n}_{e_1} = \overline{n}_{e_2}$, Eq. (2.8.27) implies two necessary conditions for instability:

$$k_z^2 > \frac{1}{3}k_\perp^2, \quad \text{and} \quad \frac{2\overline{\omega}_{pe}^2}{\Omega_e^2} > \frac{1}{3 - f} \;\; (\text{necessary for } \ell = \pm 1 \text{ instability}). \tag{2.8.28}$$

Note from Eq. (2.8.28) that the density threshold for $\ell = \pm 1$ instability is quite large, for example, $2\bar{\omega}_{pe}^2/\Omega_e^2 > 1/3$ for $f = 0$. In obtaining Eq. (2.8.28), use has been made of the definitions of $\omega_e^{\pm}$ in Eq. (2.8.13). Furthermore, $2\bar{\omega}_{pe}^2/\Omega_e^2 \leqslant 1/(1-f)$ is necessary for existence of the equilibrium [see Eq. (2.8.20)]. Denoting $\omega = \omega_r + i\omega_i$, the oscillation frequency ω_r and growth rate ω_i for the unstable branch in Eq. (2.8.25) are plotted in Fig. 2.8.5 as functions of the angle $\theta = \text{arc sin}\,(k_z/k)$, for the particular choice of parameters $\ell = 1, f = 0, \bar{n}_{e_1} = \bar{n}_{e_2}$, and $2\bar{\omega}_{pe}^2/\Omega_e^2 = 1/2$. Note that $|\omega_i|_{\max} \simeq 0.3\,\bar{\omega}_{pe}$ for the parameters used in Fig. 2.8.5, thus indicating a rather strong instability.

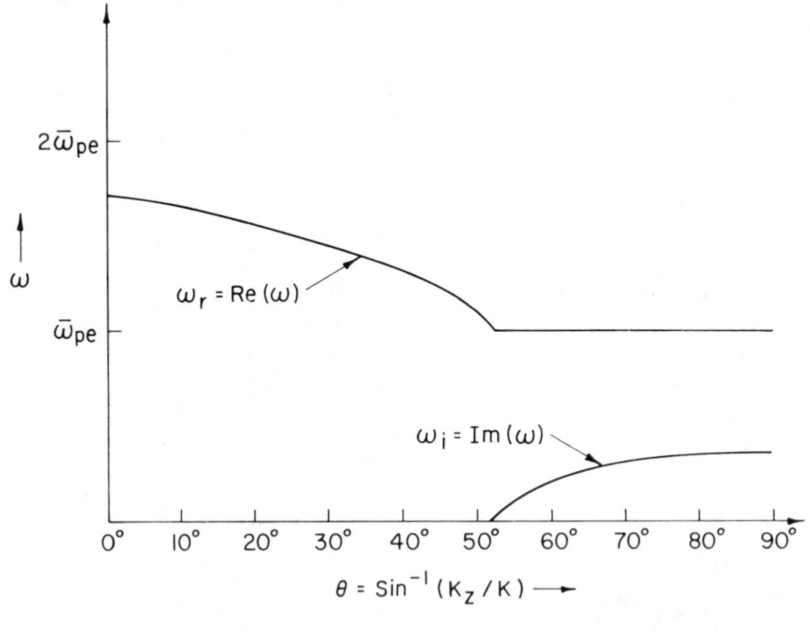

Fig. 2.8.5 Plots of oscillation frequency ω_r and growth rate ω_i versus $\theta = \text{arc sin}\,(k_z/k)$ for $\ell = 1, f = 0, \bar{n}_{e_1} = \bar{n}_{e_2}$, and $2\bar{\omega}_{pe}^2/\Omega_e^2 = 1/2$ [Eq. (2.8.25)].

For $|\ell| \geqslant 2$, and $\bar{n}_{e_1} = \bar{n}_{e_2}$, the density threshold for instability is larger than that given in Eq. (2.8.28) for $\ell = \pm 1$, but below the maximum limit for existence of the equilibrium $[2\bar{\omega}_{pe}^2/\Omega_e^2 = (1-f)^{-1}]$. Whenever the inequality in Eq. (2.8.27) is satisfied, Eq. (2.8.25) supports one unstable solution for a given value of ℓ. In the limit $k_z^2/k_\perp^2 \gg 1$, it can be verified that the maximum growth rate for a mode with azimuthal harmonic number ℓ is

$$[\omega_i]_{\max} = \{[\ell^2(\omega_e^+ - \omega_e^-)^2\, \overline{\omega}_{pe}^2/2 + \overline{\omega}_{pe}^4/4]^{1/2}$$

$$- \ell^2(\omega_e^+ - \omega_e^-)^2/4 - \overline{\omega}_{pe}^2/2\}^{1/2}, \qquad (2.8.29)$$

where $\omega_e^\pm$ are defined in Eq. (2.8.13). For $k_z^2/k_\perp^2 \gg 1$, the corresponding oscillation frequency in the unstable case is

$$\omega_r = \ell(\omega_e^+ + \omega_e^-)/2 = \ell\Omega_e/2 \qquad (2.8.30)$$

The examples discussed in Sections 2.8.2 and 2.8.3 in no way exhaust the possible waves and instabilities characteristic of nonneutral plasma-filled waveguides. For example, it is evident from Eq. (2.8.5) that there can also be instabilities associated with the relative *azimuthal* drifts of electrons and ions.† In this case the ion mass is treated as finite in Eq. (2.8.5), and only a single component of electrons is required to produce the two-rotating-stream instability. Furthermore, if there are relative *axial* drifts of plasma components (e.g., $V_{ez}^0 \neq V_{iz}^0$), Eq. (2.8.5) has unstable solutions that correspond to two-stream instabilities produced by the relative axial drifts.‡

In deriving the dispersion relation, Eq. (2.8.5), it was assumed that $R_p = R_c$. In concluding this section, it should be pointed out that Eq. (2.8.5) is also approximately correct when $R_p \neq R_c$ provided the vacuum gap between the conducting wall and the surface of the plasma column is sufficiently small, that is, $(R_c - R_p)/R_c \ll 1$. For example, if $\ell = 0$ and $k_z^2 R_c^2 \ll 1$, a careful examination of Eq. (2.7.15) shows that Eq. (2.8.5) is a valid approximation for the electrostatic dispersion relation as long as

$$p_{0m}\, \ell n(R_c/R_p) \ll 1. \qquad (2.8.31)$$

For $m = 1$, Eq. (2.8.31) reduces to

$$p_{01}\, \ell n\,(R_c/R_p) \simeq 2.4\, \ell n(R_c/R_p) \ll 1. \qquad (2.8.32)$$

---

†See Section 2.9 for a discussion of electron-ion two-rotating stream instabilities[70] in nonneutral plasma columns with $R_p < R_c$ and $2\overline{\omega}_{pe}^2/\Omega_e^2 \ll 1$.

---

‡See Section 2.11 for a discussion of two-stream instabilities produced by a relativistic electron beam drifting along the axis of a finite-radius plasma column.[59]

2.9 SURFACE WAVES ON A NONNEUTRAL PLASMA COLUMN

2.9.1 Electrostatic Dispersion Relation

In this section the electrostatic dispersion relation, Eq. (2.7.15), is examined in the limit of long axial wavelengths,[80]

$$k_z = 0. \tag{2.9.1}$$

The equilibrium configuration is illustrated in Fig. 2.7.1. It is assumed that the density of each plasma component is constant, and that the radius of the plasma column does not extend to the conducting wall, that is, $R_p < R_c$. For $k_z = 0$, Eq. (2.7.15) does not support normal-mode solutions when $\ell = 0$. Therefore the subsequent analysis in Section 2.9 is limited to perturbations with azimuthal harmonic number $\ell \neq 0$. In the limit $k_z \to 0$, Eq. (2.7.15) reduces to

$$\ell \frac{(R_p/R_c)^{2\ell} + 1}{(R_p/R_c)^{2\ell} - 1} - \ell \left[1 - \sum_\alpha \frac{\overline{\omega}_{p\alpha}^2}{(\omega - \ell\omega_\alpha)^2 - (\epsilon_\alpha \Omega_\alpha + 2\omega_\alpha)^2} \right]$$
$$= \ell \sum_\alpha \frac{\overline{\omega}_{p\alpha}^2 (\epsilon_\alpha \Omega_\alpha + 2\omega_\alpha)}{[(\omega - \ell\omega_\alpha)^2 - (\epsilon_\alpha \Omega_\alpha + 2\omega_\alpha)^2]\,(\omega - \ell\omega_\alpha)} \tag{2.9.2}$$

for $\ell \neq 0$. In Eq. (2.9.2), $\overline{\omega}_{p\alpha}^2 = 4\pi \overline{n}_\alpha e_\alpha^2/m_\alpha$, where $\overline{n}_\alpha$ = const. is the density of component α (see Fig. 2.7.1), and ω_α is the angular velocity of mean rotation defined in Eq. (2.7.3). The contributions on the right-hand side of Eq. (2.9.2) are associated with perturbations in charge density which accumulate at the surface of the plasma column. In particular, the surface terms arise from the discontinuity in $\omega_{p\alpha}^2(r)$ at $r = R_p$ [see Eq. (2.7.5) and Fig. 2.7.1]. Since the surface contributions in Eq. (2.9.2) play such an important role in determining the normal mode and stability properties of the equilibrium for $k_z = 0$, it is appropriate to refer to the $k_z = 0$ normal modes determined from Eq. (2.9.2) as *surface waves.*† Dividing Eq. (2.9.2) by ℓ and rearranging terms, we can express Eq. (2.9.2) in the compact form[80]

$$0 = 1 - \sum_\alpha \frac{\overline{\omega}_{p\alpha}^2 [1 - (R_p/R_c)^{2\ell}]}{2(\omega - \ell\omega_\alpha)[(\omega - \ell\omega_\alpha) + (\epsilon_\alpha \Omega_\alpha + 2\omega_\alpha)]}\,. \tag{2.9.3}$$

In a neutral plasma with $\Sigma_\eta\, e_\eta \overline{n}_\eta = 0$ and $\omega_\alpha = \omega_\alpha^- = 0$, Eq. (2.9.3) supports only stable solutions with Im $\omega = 0$. The situation is considerably different in a

†This terminology is further supported by the fact that the perturbed electrostatic potential $\delta\phi^\ell$ tends to be peaked around $r = R_p$ for ω satisfying Eq. (2.9.2).

nonneutral plasma. In this case the equilibrium radial electric field E_r^0 will generally produce azimuthal drift velocities that vary for different plasma components, for example, ω_α will be different for electrons and ions [see Eq. (2.7.3)]. This differential rotation can provide the free-energy source to drive two-rotating-stream instabilities for $k_z = 0$ surface waves. As in Section 2.8, no attempt will be made here to catalog all of the interesting waves and instabilities associated with Eq. (2.9.3). Rather, it is sufficient for present purposes to consider a few examples that illustrate (*a*) stable oscillations with Im $\omega = 0$ and (*b*) instability with Im $\omega > 0$.

2.9.2 Stable Oscillations

As a first application of Eq. (2.9.3) corresponding to stable oscillations, consider a neutral plasma with $\Sigma_\eta \bar{n}_\eta e_\eta = 0$ and $\omega_\alpha = \omega_\alpha^- = 0$ [see Eq. (2.7.3)]. For $\omega_\alpha = 0$, Eq. (2.9.3) reduces to

$$0 = 1 - \sum_\alpha \frac{\bar{\omega}_{p\alpha}^2 \, [1 - (R_p/R_c)^{2\ell}]}{2\omega(\omega + \epsilon_\alpha \Omega_\alpha)} . \tag{2.9.4}$$

The solutions to Eq. (2.9.4) have Im $\omega = 0$, that is, the $k_z = 0$ surface waves on a neutral plasma column do *not* exhibit instability. For a two-component neutral plasma, the exact solutions to Eq. (2.9.4) are

$$\omega = 0,$$

$$\omega = \frac{\Omega_e - \Omega_i}{2} \pm \frac{1}{2} \left\{ (\Omega_e - \Omega_i)^2 + 4\Omega_e\Omega_i + 2(\bar{\omega}_{pe}^2 + \bar{\omega}_{pi}^2) \left[1 - \left(\frac{R_p}{R_c} \right)^{2\ell} \right] \right\}^{1/2} , \tag{2.9.5}$$

where $\Omega_e \equiv eB_0/m_e c$ and $\Omega_i \equiv e_i B_0/m_i c$ are the electron and ion cyclotron frequencies, respectively.

As a second application of Eq. (2.9.3) corresponding to stable oscillations, consider a nonneutral plasma with no ions ($\bar{n}_i = 0$), and a *single* component of electrons rotating with angular velocity $\omega_\alpha = \omega_e^+$ or $\omega_\alpha = \omega_e^-$, where

$$\omega_e^\pm = \frac{\Omega_e}{2} \left\{ 1 \pm \left[1 - \frac{2\bar{\omega}_{pe}^2}{\Omega_e^2} \right]^{1/2} \right\} . \tag{2.9.6}$$

In this case Eq. (2.9.3) reduces to

$$0 = 1 - \frac{\overline{\omega}_{pe}^2 \, [1 - (R_p/R_c)^{2\ell}]}{2(\omega - \ell\omega_e)\,[(\omega - \ell\omega_e) + (-\Omega_e + 2\omega_e)]} \cdot \tag{2.9.7}$$

The solutions to Eq. (2.9.7) have Im $\omega = 0$, that is, the $k_z = 0$ surface waves on a nonneutral plasma column consisting of a single component of electrons do *not* exhibit instability. The exact solutions to Eq. (2.9.7) are

$$\omega = \ell\omega_e + \left(\frac{\Omega_e}{2} - \omega_e\right) \pm \left\{\left(\frac{\Omega_e}{2} - \omega_e\right)^2 + \frac{\overline{\omega}_{pe}^2}{2}\left[1 - \left(\frac{R_p}{R_c}\right)^{2\ell}\right]\right\}^{1/2}. \tag{2.9.8}$$

In the limit where the conducting wall is located at infinity $R_c \to \infty$, it follows from Eqs. (2.9.6) and (2.9.8) that the normal mode frequencies are

$$\omega = (\ell - 1)\omega_e + \Omega_e, \qquad \text{and} \qquad \omega = (\ell - 1)\omega_e. \tag{2.9.9}$$

2.9.3 Electron-Ion Two-Rotating-Stream Instability

As an application of Eq. (2.9.3) corresponding to instability, consider a two-component nonneutral plasma consisting of electrons and ions. The ions and electron densities are related by

$$\overline{n}_i = f\overline{n}_e, \tag{2.9.10}$$

where f = const. = fractional neutralization. Single ionization is assumed. It is readily verified from Eq. (2.7.3) that the two allowed values of equilibrium rotation velocity are

$$\omega_e = \omega_e^{\pm} = \frac{\Omega_e}{2}\left\{1 \pm \left[1 - \frac{2\overline{\omega}_{pe}^2}{\Omega_e^2}(1 - f)\right]^{1/2}\right\} \tag{2.9.11}$$

for the electrons, and

$$\omega_i = \omega_i^{\pm} = -\frac{\Omega_i}{2}\left\{1 \pm \left[1 + \frac{2\overline{\omega}_{pe}^2}{\Omega_e^2}\frac{m_i}{m_e}(1 - f)\right]^{1/2}\right\} \tag{2.9.12}$$

for the ions. It is assumed that the equilibrium consists of a *single* component of electrons rotating with angular velocity $\omega_e = \omega_e^-$ *or* $\omega_e = \omega_e^+$, and a *single* component of ions rotating with angular velocity $\omega_i = \omega_i^-$ *or* $\omega_i = \omega_i^+$. For purposes of the stability analysis, it is not necessary to prescribe whether the electrons or ions are in the fast rotational ($\omega_\alpha = \omega_\alpha^+$) or the slow rotational ($\omega_\alpha = \omega_\alpha^-$) mode.

Since ω_i and ω_e are generally different, it can be anticipated that Eq. (2.9.3) supports unstable solutions with Im $\omega > 0$, corresponding to a two-stream instability produced by the differential rotation of electrons and ions. For present purposes, the analysis of Eq. (2.9.3) is limited to the fundamental ($\ell = 1$) mode. Substituting $\ell = 1$ in Eq. (2.9.3), it is straightforward to show that Eq. (2.9.3) can be expressed as

$$\frac{1}{1-(R_p/R_c)^2} = \frac{\overline{\omega}_{pe}^2}{2(\omega-\omega_e^-)(\omega-\omega_e^+)} + \frac{\overline{\omega}_{pi}^2}{2(\omega-\omega_i^-)(\omega-\omega_i^+)}, \tag{2.9.13}$$

where $\omega_e^\pm$ and $\omega_i^\pm$ are defined in Eqs. (2.9.11) and (2.9.12). For a neutral plasma with $f = 1$, the solutions to Eq. (2.9.13) reduce to Eq. (2.9.5) for $\ell = 1$, and there is no instability. Furthermore, for a pure electron gas with $f = 0$, the solutions to Eq. (2.9.13) reduce to Eq. (2.9.8) for $\ell = 1$, and there is no instability. For $f \neq 0$ and $f \neq 1$, however, Eq. (2.9.13) has an unstable root with Im $\omega > 0$, at least in certain parameter regimes.

To simplify the stability analysis of Eq. (2.9.13), consider a *low-density* nonneutral plasma with

$$\frac{2\overline{\omega}_{pe}^2}{\Omega_e^2}(1-f) \ll 1. \tag{2.9.14}$$

Making use of Eqs. (2.9.11) and (2.9.14), we can express $\omega_e^\pm$ in the approximate forms

$$\omega_e^+ = \Omega_e \tag{2.9.15}$$

and

$$\omega_e^- = \frac{\overline{\omega}_{pe}^2}{2\Omega_e}(1-f). \tag{2.9.16}$$

For low densities, note from Eq. (2.9.16) that ω_e^- is the $\mathbf{E}^0 \times \mathbf{B}_0$ rotation velocity, that is, $\omega_e^- \simeq -cE_r^0(r)/rB_0$ for $r \leqslant R_p$. For frequencies well below the electron cyclotron frequency,

$$|\omega| \ll \Omega_e, \tag{2.9.17}$$

Eq. (2.9.13) reduces to

$$\frac{1}{1-(R_p/R_c)^2} = -\frac{\overline{\omega}_{pe}^2/2\Omega_e}{\omega-\omega_e^-} + \frac{\overline{\omega}_{pi}^2}{2(\omega-\omega_i^-)(\omega-\omega_i^+)}, \tag{2.9.18}$$

where $\omega_i^\pm$ and ω_e^- are defined in Eqs. (2.9.12) and (2.9.16). Equation (2.9.18) was first derived by Levy, Daugherty, and Buneman[70] and is identical to their Eq. (58). For a pure electron gas with $f=0$, the solution to Eq. (2.9.18) is

$$\omega = \Omega_D \equiv \omega_e^-(R_p/R_c)^2. \tag{2.9.19}$$

If a low-density ion component is present ($0<f\ll 1$), Eq. (2.9.18) has one unstable solution with Im $\omega>0$ provided $\Omega_D \simeq \omega_i^-$, that is, provided there is a *resonance* between the electron diocotron wave and the azimuthal motion of the ions.† For $f\ll 1$, it can be shown that instability exists only if Ω_D and ω_i^- differ by an amount proportional to $\sqrt{f}$. Furthermore, for $f\ll 1$ and Ω_D very close to ω_i^-, it can be shown that the growth rate is approximately[70]

$$\omega_i = \text{Im}\,\omega = \overline{\omega}_{pi}\left\{\left(1-\frac{R_p^2}{R_c^2}\right)^3\left(\frac{R_c^2/2R_p^2}{2-R_p^2/R_c^2}\right)\right\}^2. \tag{2.9.20}$$

Note from Eq. (2.9.20) that the growth rate is of order $\overline{\omega}_{pi}$ times a geometric factor.

Since Eq. (2.9.18) is a cubic equation for ω, it can be solved exactly for the complex oscillation frequency as a function of the parameters f, $(R_p/R_c)^2$, ω_e^-, and so on. It is adequate for present purposes, however, to determine the range of system parameters for which there is instability (Im $\omega>0$). Shown in Fig. 2.9.1 are stability-instability boundaries for

$$f \equiv \frac{\overline{n}_i}{\overline{n}_e} \text{ versus } \lambda \equiv \frac{m_e}{m_i}\,\frac{2\Omega_e^2/\overline{\omega}_{pe}^2}{1-f},$$

for three different values of R_p/R_c. For a given value of R_p/R_c, Eq. (2.9.18) gives pure oscillatory solutions (Im $\omega=0$) for values of f and λ falling *below* the curve. For values of f and λ *above* the curve, Eq. (2.9.18) has one unstable solution with Im $\omega>0$. Evidently, if $f=0$ or $f=1$ ($\lambda=\infty$), there is no instability. Furthermore, for given values of f and λ, the instability is stabilized if the conducting wall is sufficiently close to the plasma column, that is, if R_p/R_c is sufficiently close to unity.

†Note from Eqs. (2.9.12) and (2.9.19) that $\omega_i^+<0$ whereas $\omega_i^->0$ and $\Omega_D>0$ (for $B_0>0$).

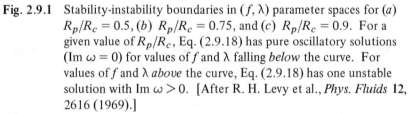

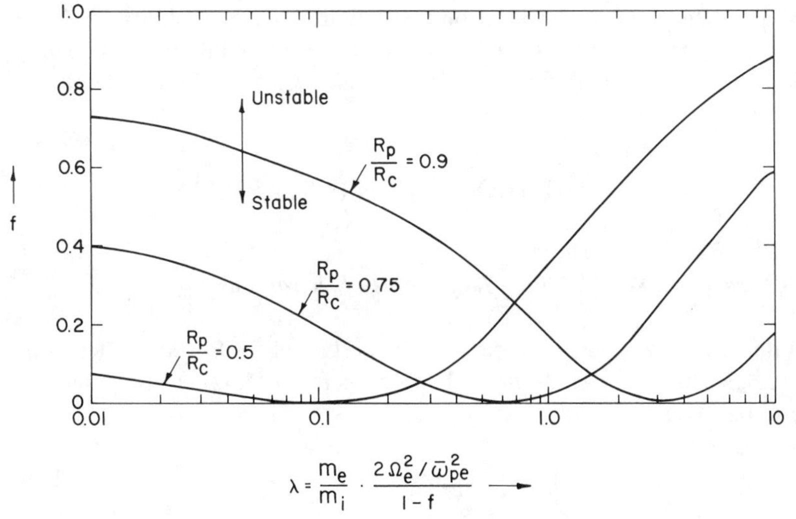

Fig. 2.9.1 Stability-instability boundaries in (f, λ) parameter spaces for (*a*) $R_p/R_c = 0.5$, (*b*) $R_p/R_c = 0.75$, and (*c*) $R_p/R_c = 0.9$. For a given value of R_p/R_c, Eq. (2.9.18) has pure oscillatory solutions ($\mathrm{Im}\,\omega = 0$) for values of f and λ falling *below* the curve. For values of f and λ *above* the curve, Eq. (2.9.18) has one unstable solution with $\mathrm{Im}\,\omega > 0$. [After R. H. Levy et al., *Phys. Fluids* **12**, 2616 (1969).]

One of the most important features of the electron-ion two-rotating-stream instability is that it can exist for *low* densities [Eq. (2.9.14)] and for small values of $f \ll 1$. There is no density threshold for instability, as is the case for the electron-electron two-rotating-stream instability in a nonneutral plasma-filled waveguide (Section 2.8.3). Note also that the characteristic growth rate of the instability is substantial [Eq. (2.9.20)].

In concluding this section, it is important to note that the present discussion of two-rotating-stream instabilities for $k_z = 0$ surface waves is by no means complete. For example, Eq. (2.9.13) also predicts instability if the assumptions of low density [Eq. (2.9.14)] and low frequency [Eq. (2.9.17)] are relaxed. Furthermore, the electron-ion two-rotating-stream instability can also exist for $\ell \geq 2$.[70] The corresponding growth rates, however, are smaller than those for the fundamental ($\ell = 1$) mode. Finally, if no ions are present in the system ($f = 0$), but two components of electrons are rotating with different angular velocities ω_e^+ and ω_e^- [Eq. (2.9.6)], then Eq. (2.9.3) can have unstable solutions corresponding to an electron-electron two-rotating-stream instability.[80] Although the growth rates are much larger for the electron-electron instability $[(\mathrm{Im}\ \omega)_{max} \approx \Omega_e]$, the density threshold is also larger ($2\bar{\omega}_{pe}^2/\Omega_e^2 > 3/4$, for equidensity electron components).

2.10 THE DIOCOTRON INSTABILITY

2.10.1 Poisson's Equation for a Low-Density Electron Gas

In this section, Poisson's equation for the perturbed electrostatic potential, Eq. (2.6.20), is used to study the stability properties of low-density, pure electron gas equilibria with

$$\omega_{pe}^2(r) \ll \Omega_e^2, \tag{2.10.1}$$

and

$$n_i^0(r) = 0, \tag{2.10.2}$$

where $\omega_{pe}^2(r) = 4\pi n_e^0(r)e^2/m_e$, and $\Omega_e = eB_0/m_e c$. It is assumed that the axial wave vector of the perturbation is equal to zero,

$$k_z = 0, \tag{2.10.3}$$

and that the frequency of the perturbation is low with

$$|\omega - \ell\omega_e(r)|^2 \ll \Omega_e^2\,. \tag{2.10.4}$$

Consistent with Eqs. (2.10.1) and (2.10.4), it is also assumed that the electron column is rotating in the slow rotational mode with angular velocity

$$\omega_e(r) = \omega_e^-(r). \tag{2.10.5}$$

For low densities, $\omega_e^-(r)$ can be expressed in the approximate form [see Eq. 2.2.8)]

$$\omega_e^-(r) = \frac{1}{r^2\Omega_e}\int_0^r dr'\, r'\omega_{pe}^2(r') = -\frac{cE_r^0(r)}{rB_0}, \tag{2.10.6}$$

where $E_r^0(r)$ is the equilibrium radial electric field [see Eq. (2.1.2)]. Note from Eq. (2.10.6) that the mean motion of the electron column corresponds to an $\mathbf{E}^0 \times \mathbf{B}_0$ rotation about the axis of symmetry.

Making use of Eqs. (2.10.1)-(2.10.5), we can express Poisson's equation for the perturbed electrostatic potential, Eq. (2.6.20), in the approximate form[106]

$$\frac{1}{r}\frac{\partial}{\partial r}r\frac{\partial}{\partial r}\delta\phi^{\ell} - \frac{\ell^2}{r^2}\delta\phi^{\ell} = -\frac{\ell}{\Omega_e r}\,\frac{\delta\phi^{\ell}}{\omega - \ell\omega_e^-(r)}\,\frac{\partial}{\partial r}\omega_{pe}^2(r), \tag{2.10.7}$$

where Im $\omega > 0$,† and $\delta\phi^{\ell} \equiv \delta\phi^{\ell}(r, k_z = 0)$. In Eq. (2.10.7), $\omega_{pe}^2(r)$ and $\omega_e^-(r)$ are arbitrary functions of r consistent with Eq. (2.10.6) and $\omega_{pe}^2(r) \geqslant 0$. Without loss of generality, we assume that $\ell \geqslant 1$ throughout the remainder of Section 2.10.

2.10.2 A Sufficient Condition for Stability

It is straightforward to show from Eq. (2.10.7) that the equilibrium configuration is stable if $\partial\omega_{pe}^2(r)/\partial r \leqslant 0$.[110] In other words, Eq. (2.10.7) supports no unstable solutions with Im $\omega > 0$‡ if $\omega_{pe}^2(r)$ is a monotonically decreasing function of r,

$$\frac{\partial}{\partial r}\omega_{pe}^2(r) \leqslant 0. \tag{2.10.8}$$

It is assumed that a grounded conducting wall is located at $r = R_c$. Multiplying Eq. (2.10.7) by $r(\delta\phi^{\ell})^*$ [where $(\delta\phi^{\ell})^*$ is the complex conjugate of $\delta\phi^{\ell}$] and integrating from $r = 0$ to $r = R_c$ gives

$$0 = D(\omega) \equiv \int_0^{R_c} dr\, r\left\{\left|\frac{\partial}{\partial r}\delta\phi^{\ell}\right|^2 + \frac{\ell^2}{r^2}\left|\delta\phi^{\ell}\right|^2\right\}$$

$$-\frac{\ell}{\Omega_e}\int_0^{R_c} dr\,\frac{|\delta\phi^{\ell}|^2}{\omega - \ell\omega_e^-(r)}\,\frac{\partial}{\partial r}\omega_{pe}^2(r), \tag{2.10.9}$$

where Im $\omega > 0$, and the boundary conditions, $[\delta\phi^{\ell}]_{r=R_c} = 0$ and $[r\partial\delta\phi^{\ell}/\partial r]_{r=0} = 0$, have been enforced in obtaining Eq. (2.10.9). In Eq. (2.10.9) the complex oscillation frequency can be expressed as

$$\omega = \omega_r + i\omega_i, \tag{2.10.10}$$

†The dispersion relation can be defined in the region Im $\omega \leqslant 0$ by appropriate analytic continuation.

‡Keep in mind that the time variation in perturbed quantities is assumed to be proportional to exp $(-i\omega t)$. Therefore, Im $\omega = \omega_i > 0$ corresponds to instability.

where ω_r amd ω_i are real. Equating to zero the real and imaginary parts of Eq. (2.10.9) gives

$$0 = \int_0^{R_c} dr\, r \left\{ \left| \frac{\partial}{\partial r} \delta\phi^\ell \right|^2 + \frac{\ell^2}{r^2} \left| \delta\phi^\ell \right|^2 \right\}$$

$$- \frac{\ell}{\Omega_e} \int_0^{R_c} dr \frac{|\delta\phi^\ell|^2 [\omega_r - \ell\omega_e(r)]}{[\omega_r - \ell\omega_e^-(r)]^2 + \omega_i^2} \frac{\partial}{\partial r} \omega_{pe}^2(r), \qquad (2.10.11)$$

$$0 = \omega_i \frac{\ell}{\Omega_e} \int_0^{R_c} dr \frac{|\delta\phi^\ell|^2}{[\omega_r - \ell\omega_e^-(r)]^2 + \omega_i^2} \frac{\partial}{\partial r} \omega_{pe}^2(r). \qquad (2.10.12)$$

If $\partial\omega_{pe}^2(r)/\partial r \leqslant 0$ over the interval $0 < r < R_c$, the integral in Eq. (2.10.12) is negative. Therefore it follows from Eq. (2.10.12) that the dispersion relation has *no* unstable solutions with $\text{Im}\ \omega = \omega_i > 0$ if $\omega_{pe}^2(r)$ is a monotonically decreasing function of r.†

Equation (2.10.8) is a *sufficient* condition for *stability* of the equilibrium configuration. Therefore a *necessary* condition for *instability* can be stated as follows:

$$\frac{\partial}{\partial r} \omega_{pe}^2(r) \text{ changes sign on the interval } 0 < r < R_c. \qquad (2.10.13)$$

In other words, Eq. (2.10.7) has unstable solutions with $\text{Im}\ \omega = \omega_i > 0$ only if the radial density gradient changes sign on the interval $0 < r < R_c$, for example, if $\omega_{pe}^2(r)$ corresponds to a hollow electron beam equilibrium. It should be emphasized that Eq. (2.10.13) is a necessary condition, but not a sufficient condition, for instability. Not all equilibrium configurations in which $\partial\omega_{pe}^2(r)/\partial r$ changes sign are unstable. Making use of the definition of $\omega_e^-(r)$ in terms of $\omega_{pe}^2(r)$, we can state the necessary condition for instability, Eq. (2.10.13), in the equivalent form

†Similarly, if $\omega_{pe}^2(r)$ is a monotonically increasing function of r with $\partial\omega_{pe}^2(r)/\partial r \geqslant 0$ over the interval $0 < r < R_c$, the dispersion relation, Eq. (2.10.9), has no unstable solutions with $\text{Im}\ \omega = \omega_i > 0$.

$$\frac{\partial}{\partial r}\frac{1}{r}\frac{\partial}{\partial r}[r^2\omega_e^-(r)] \text{ changes sign on the interval } 0<r<R_c. \tag{2.10.14}$$

It is evident from Eq. (2.10.14) that a shear in angular velocity $\omega_e^-(r)$ is required for instability.

2.10.3 Stability of Hollow Beam Equilibria

As an example of an equilibrium configuration that exhibits the diocotron (slipping stream) instability,[98–111] consider the hollow electron beam equilibrium illustrated in Fig. 2.10.1.

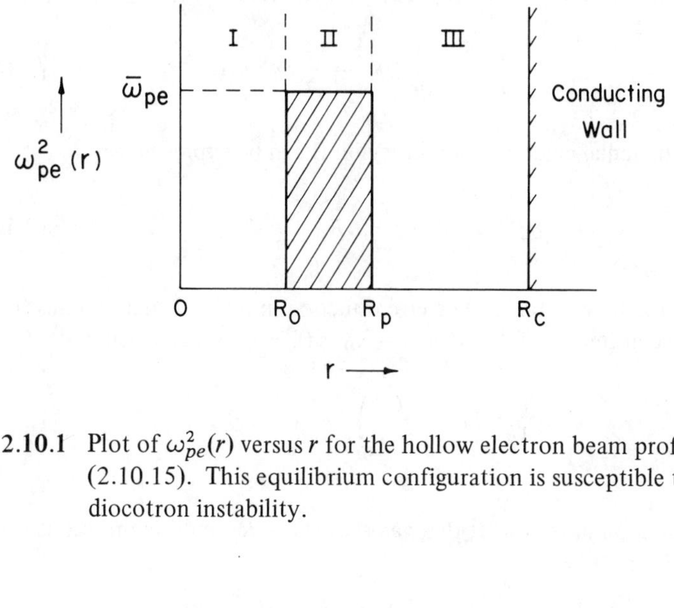

Fig. 2.10.1 Plot of $\omega_{pe}^2(r)$ versus r for the hollow electron beam profile in Eq. (2.10.15). This equilibrium configuration is susceptible to the diocotron instability.

The density profile has the form

$$\omega_{pe}^2(r) = \begin{cases} 0, & 0<r<R_0 \quad \text{(region I)}, \\ \bar{\omega}_{pe}^2, & R_0<r<R_p \quad \text{(region II)}, \\ 0, & R_p<r<R_c \quad \text{(region III)}, \end{cases} \tag{2.10.15}$$

where $\bar{\omega}_{pe}^2 = 4\pi\bar{n}_e e^2/m_e = \text{const.}$ Evidently, for the density profile given in Eq. (2.10.15), $\partial\omega_{pe}^2(r)/\partial r$ can be expressed as

$$\frac{\partial}{\partial r}\omega_{pe}^2(r) = \bar{\omega}_{pe}^2[\delta(r-R_0)-\delta(r-R_p)]\,. \tag{2.10.16}$$

Moreover, it follows from Eqs. (2.10.6) and (2.10.15) that the angular velocity profile is given by [see also Eq. (2.2.16)]

$$\omega_e^-(r) = \frac{\bar{\omega}_{pe}^2}{2\Omega_e}\left(1-\frac{R_0^2}{r^2}\right), \quad R_0 < r < R_p. \tag{2.10.17}$$

Note from Eq. (2.10.17) that $\partial\omega_e^-(r)/\partial r \neq 0$ within the beam, that is, there is a *shear* in angular velocity of mean rotation provided $R_0 \neq 0$.

Except at $r = R_0$ and $r = R_p$, Poisson's equation for the perturbed electrostatic potential reduces to [see Eqs. (2.10.7) and (2.10.16)]

$$\frac{1}{r}\frac{\partial}{\partial r}r\frac{\partial}{\partial r}\delta\phi^{\ell} - \frac{\ell^2}{r^2}\delta\phi^{\ell} = 0. \tag{2.10.18}$$

Therefore the radial eigenfunction in region II can be expressed as

$$\delta\phi_{\mathrm{II}}^{\ell} = Br^{\ell} + \frac{C}{r^{\ell}}, \quad R_0 < r < R_p, \tag{2.10.19}$$

where B and C are constants. The eigenfunction in region I that remains finite at $r = 0$ and is continuous with $\delta\phi_{\mathrm{II}}^{\ell}$ at $r = R_0$ is ($\ell \geqslant 1$ has been assumed)

$$\delta\phi_{\mathrm{I}}^{\ell} = \left(B + \frac{C}{R_0^{2\ell}}\right)r^{\ell}, \quad 0 < r < R_0. \tag{2.10.20}$$

The eigenfunction in region III that vanishes at $r = R_c$ and is continuous with $\delta\phi_{\mathrm{II}}^{\ell}$ at $r = R_p$ is

$$\delta\phi_{\mathrm{III}}^{\ell} = \left(BR_p^{2\ell} + C\right)\frac{R_c^{2\ell} - r^{2\ell}}{R_c^{2\ell} - R_p^{2\ell}}\frac{1}{r^{\ell}}. \tag{2.10.21}$$

The constants B and C in Eqs. (2.10.19)-(2.10.21) can be related by integrating Eq. (2.10.7) across the discontinuities at $r = R_0$ and $r = R_p$. Multiplying Eq. (2.10.7) by r, integrating from $r = R_p - \epsilon$ to $r = R_p + \epsilon$, and taking the limit $\epsilon \to 0_+$ gives

$$R_p \left[\frac{\partial}{\partial r} \delta\phi_{\text{III}}^{\ell} \right]_{r=R_p} - R_p \left[\frac{\partial}{\partial r} \delta\phi_{\text{II}}^{\ell} \right]_{r=R_p}$$

$$= \ell [\delta\phi^{\ell}]_{r=R_p} \frac{\bar{\omega}_{pe}^2}{\Omega_e [\omega - \ell \omega_e^- (R_p)]} . \tag{2.10.22}$$

Equation (2.10.22) relates the discontinuity in radial electric field at $r = R_p$ to the perturbation in surface charge density. In a similar manner, multiplying Eq. (2.10.7) by r, integrating from $r = R_0 - \epsilon$ to $r = R_0 + \epsilon$, and taking the limit $\epsilon \to 0_+$ gives

$$R_0 \left[\frac{\partial}{\partial r} \delta\phi_{\text{II}}^{\ell} \right]_{r=R_0} - R_0 \left[\frac{\partial}{\partial r} \delta\phi_{\text{I}}^{\ell} \right]_{r=R_0}$$

$$= -\ell [\delta\phi^{\ell}]_{r=R_0} \frac{\bar{\omega}_{pe}^2}{\Omega_e [\omega - \ell \omega_e^- (R_0)]} . \tag{2.10.23}$$

Making use of Eq. (2.10.17), we can express the rotation frequencies that occur in Eqs. (2.10.22) and (2.10.23) as

$$\omega_e^- (R_p) = \omega_{\text{D}} \left(1 - \frac{R_0^2}{R_p^2} \right) , \tag{2.10.24}$$

$$\omega_e^- (R_0) = 0, \tag{2.10.25}$$

where ω_{D} is defined by†

$$\omega_{\text{D}} \equiv \frac{\bar{\omega}_{pe}^2}{2\Omega_e} . \tag{2.10.26}$$

Substituting Eqs. (2.10.19)-(2.10.21) into Eqs. (2.10.22) and (2.10.23) and making use of Eqs. (2.10.24)-(2.10.26) results in the following eigenvalue equation for ω:[106]

$$(\omega / \omega_{\text{D}})^2 - b(\omega / \omega_{\text{D}}) + c = 0, \tag{2.10.27}$$

where

$$b \equiv \ell \left(1 - \frac{R_0^2}{R_p^2} \right) + \left(\frac{R_p^{2\ell}}{R_c^{2\ell}} - \frac{R_0^{2\ell}}{R_c^{2\ell}} \right) , \tag{2.10.28}$$

$$c \equiv \ell \left(1 - \frac{R_0^2}{R_p^2} \right) \left(1 - \frac{R_0^{2\ell}}{R_c^{2\ell}} \right) - \left(1 - \frac{R_0^{2\ell}}{R_p^{2\ell}} \right) \left(1 - \frac{R_p^{2\ell}}{R_c^{2\ell}} \right) . \tag{2.10.29}$$

Equation (2.10.27) is a quadratic equation for the complex eigenfrequency ω. Its solution is

---

†Note from Eqs. (2.10.17) and (2.10.26) that ω_{D} is the $\mathbf{E}^0 \times \mathbf{B}_0$ rotation velocity for a solid electron beam ($R_0 = 0$).

$$\omega = \frac{\omega_{\mathrm{D}}}{2}\left[b \pm (b^2 - 4c)^{1/2}\right]. \qquad (2.10.30)$$

If $b^2 \geqslant 4c$, the oscillation frequencies defined in Eq. (2.10.30) are real ($\mathrm{Im}\,\omega = 0$), and the equilibrium configuration is not unstable. From Eqs. (2.10.28) and (2.10.29), the condition for stability ($b^2 \geqslant 4c$) can be expressed in the equivalent form

$$\left[-\ell\left(1 - \frac{R_0^2}{R_p^2}\right)R_c^{2\ell} + 2R_c^{2\ell} - R_p^{2\ell} - R_0^{2\ell}\right]^2$$
$$\geqslant 4R_0^{2\ell}R_p^{-2\ell}\left(R_c^{2\ell} - R_p^{2\ell}\right)^2 . \qquad (2.10.31)$$

Evidently, if $R_0 = 0$ (which corresponds to a solid electron beam), or $R_p = R_c$ (which corresponds to a hollow electron beam with outer radius extending to the conducting wall), the inequality in Eq. (2.10.31) is trivially satisfied, and there is *no* instability (see Section 2.10.2).

If the inequality in Eq. (2.10.31) is violated ($b^2 < 4c$), the oscillation frequencies in Eq. (2.10.30) are complex and form conjugate pairs. Defining $\omega = \omega_r + i\omega_i$, where ω_r and ω_i are real, we obtain for Eq. (2.10.30)

$$\omega_r = \frac{1}{2}b\omega_{\mathrm{D}}, \qquad (2.10.32)$$

$$\omega_i = \pm\frac{1}{2}(4c - b^2)^{1/2}\,\omega_{\mathrm{D}}, \qquad (2.10.33)$$

where $4c > b^2$ is assumed. The solution with $\mathrm{Im}\,\omega = \omega_i > 0$ in Eq. (2.10.33) corresponds to instability. Note from Eqs. (2.10.32) and (2.10.33) that ω_r and ω_i are each of order ω_{D} times a geometric factor.

For $\ell = 1$, it is important to note that Eq. (2.10.31) reduces to

$$(R_c^2 - R_p^2)^2\,(R_p^2 - R_0^2)^2 \geqslant 0. \qquad (2.10.34)$$

Evidently, Eq. (2.10.34) is always satisfied, and the system does not exhibit instability for $\ell = 1$. For $\ell = 2$, Eq. (2.10.31) can be expressed as

$$R_p^2 + R_0^2 \geqslant 2\frac{R_0}{R_c}\frac{R_c}{R_p}R_c^2 . \qquad (2.10.35)$$

The inequality in Eq. (2.10.35) is violated for sufficiently large R_0/R_c and/or small R_p/R_c, in which case the system is unstable for $\ell = 2$. Equation (2.10.31) can be used to obtain stability-instability curves as functions of the parameters

R_p/R_c and R_0/R_c, for various values of harmonic number ℓ. Shown in Fig. 2.10.2 are the stability-instability curves for $\ell = 2, 3, 4$. Only the region above

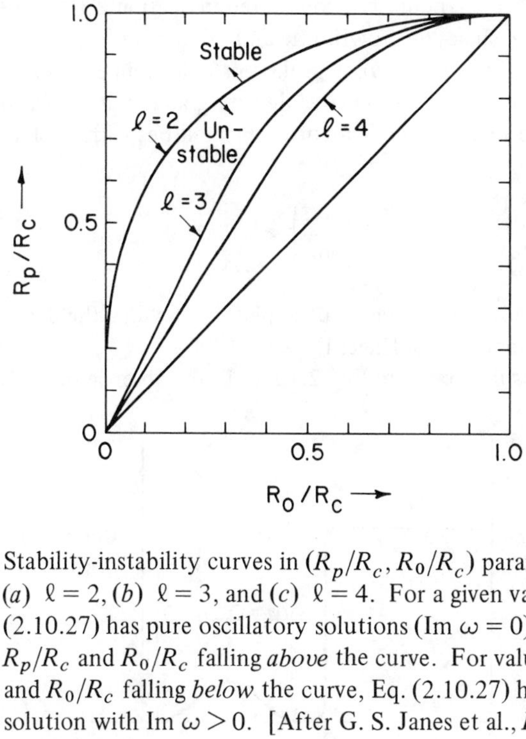

Fig. 2.10.2 Stability-instability curves in $(R_p/R_c, R_0/R_c)$ parameter space for (*a*) $\ell = 2$, (*b*) $\ell = 3$, and (*c*) $\ell = 4$. For a given value of ℓ, Eq. (2.10.27) has pure oscillatory solutions ($\text{Im}\,\omega = 0$) for values of R_p/R_c and R_0/R_c falling *above* the curve. For values of R_p/R_c and R_0/R_c falling *below* the curve, Eq. (2.10.27) has one unstable solution with $\text{Im}\,\omega > 0$. [After G. S. Janes et al., *Phys. Rev.* **145**, 925 (1966).]

the 45° line in Fig. 2.10.2 has meaning, since $R_p > R_0$. For a given value of ℓ, the region of parameter space above the curve corresponds to stability ($\text{Im}\,\omega = 0$), whereas the region between the curve and the 45° line corresponds to instability ($\text{Im}\,\omega > 0$). Evidently, for a fixed value of R_0/R_c, a sufficiently large value of R_p/R_c assures stability. Alternatively, for a fixed value of R_p/R_c, a sufficiently small value of R_0/R_c assures stability. In the nonlinear regime, it is reasonable to speculate that the diocotron instability is stabilized by a radial diffusion of the electron density profile, which results in an increase in the effective value of R_p and/or a decrease in the effective value of R_0. In other words, if the nonlinear response of the electron density profile to the unstable field perturbations is calculated, it can be anticipated that the density profile adjusts to alleviate the instability.

2.10.4 Resonant Diocotron Instability

For the hollow beam equilibrium considered in Section 2.10.3, the diocotron instability is *strong*, since the growth rate and oscillation frequency are comparable in magnitude [i.e., $\omega_i = 0(\omega_r)$] in the unstable regime [see Eqs. (2.10.32) and (2.10.33)]. This is due to the large density gradient at $r = R_0$ that drives the instability (see Fig. 2.10.1). Diocotron instabilities with small growth rates,

$$\left|\frac{\omega_i}{\omega_r}\right| \ll 1, \tag{2.10.36}$$

can also exist in a low-density nonneutral plasma provided the density gradient that drives the instability is sufficiently weak.[110]

A case in point is illustrated in Fig. 2.10.3. In the absence of the density

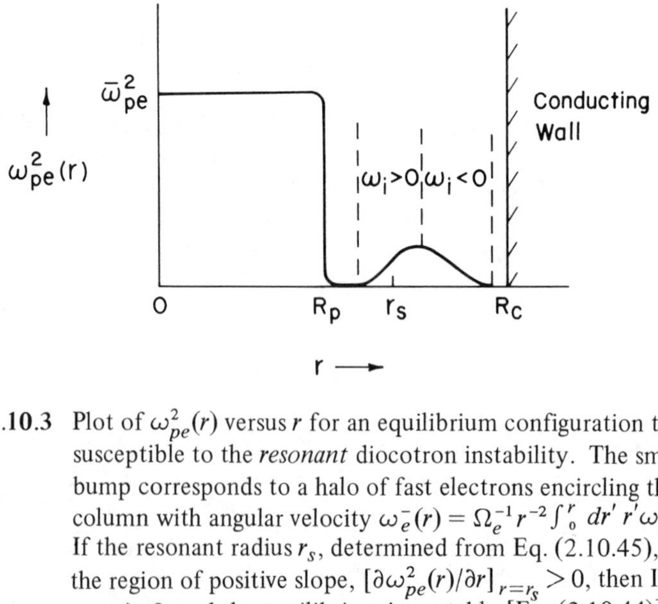

Fig. 2.10.3 Plot of $\omega^2_{pe}(r)$ versus r for an equilibrium configuration that is susceptible to the *resonant* diocotron instability. The small density bump corresponds to a halo of fast electrons encircling the main column with angular velocity $\omega_e^-(r) = \Omega_e^{-1} r^{-2} \int_0^r dr'\, r' \omega^2_{pe}(r')$. If the resonant radius r_s, determined from Eq. (2.10.45), falls in the region of positive slope, $[\partial\omega^2_{pe}(r)/\partial r]_{r=r_s} > 0$, then Im $\omega = \omega_i > 0$, and the equilibrium is unstable [Eq. (2.10.44)].

bump in the interval $R_p < r < R_c$, the equilibrium configuration in Fig. 2.10.3 is stable with $\omega_i = 0$ and

$$\omega_r = \omega_D[\ell - 1 + (R_p/R_c)^{2\ell}]. \tag{2.10.37}$$

Equation (2.10.37) follows directly from Eqs. (2.10.27)-(2.10.29) with $R_0 = 0$. The mode in Eq. (2.10.37) corresponds to a surface wave associated with the discontinuity in density at the surface ($r = R_p$) of the main plasma column. The addition of the density bump in Fig. 2.10.3 leads to the possibility of instability since $\partial\omega_{pe}^2(r)/\partial r$ changes sign in the interval $R_p < r < R_c$. [Note that this small density bump corresponds to a halo of fast electrons encircling the main column in Fig. 2.10.3 with angular velocity $\omega_e^-(r) = \Omega_e^{-1} r^{-2} \int_0^r dr'\, r' \omega_{pe}^2(r')$.] Since the density bump is small and the corresponding spatial gradients are weak, it is anticipated that the growth rate ω_i is small and that the oscillation frequency ω_r is given by Eq. (2.10.37) to a good approximation.

The procedure for determining ω_i in circumstances where $|\omega_i/\omega_r| \ll 1$ can be summarized is as follows. The right-hand side of Eq. (2.10.9) is denoted by $D(\omega) = D(\omega_r + i\omega_i)$. Taylor-expanding $D(\omega_r + i\omega_i)$ for small ω_i, and dividing $D(\omega_r)$ into its real and imaginary parts according to $D(\omega_r) = D_r(\omega_r) + iD_i(\omega_r)$, we can express Eq. (2.10.9) as

$$0 = \left[D_r(\omega_r) - \omega_i \frac{\partial}{\partial\omega_r} D_i(\omega_r)\right] + i\left[D_i(\omega_r) + \omega_i \frac{\partial}{\partial\omega_r} D_r(\omega_r)\right] + \dots, \tag{2.10.38}$$

where

$$D_r(\omega_r) \equiv \int_0^{R_c} dr \left\{ r \left| \frac{\partial}{\partial r} \delta\phi^\ell \right|^2 + \frac{\ell^2}{r} |\delta\phi^\ell|^2 - \frac{\ell |\delta\phi^\ell|^2}{\Omega_e} \frac{P}{\omega_r - \ell\omega_e^-(r)} \frac{\partial}{\partial r} \omega_{pe}^2(r) \right\} \tag{2.10.39}$$

and

$$D_i(\omega_r) \equiv \frac{\pi\ell}{\Omega_e} \int_0^{R_c} dr\, |\delta\phi^\ell|^2\, \delta[\omega_r - \ell\omega_e^-(r)] \frac{\partial}{\partial r} \omega_{pe}^2(r). \tag{2.10.40}$$

In Eq. (2.10.39), P denotes the Cauchy principal value. In obtaining Eqs. (2.10.39) and (2.10.40) use has been made of the Plemelj formula,

$$\lim_{\delta \to 0_+} \frac{1}{\omega_r - \ell\omega_e^-(r) + i\delta} \equiv \frac{P}{\omega_r - \ell\omega_e^-(r)} - i\pi\,\delta[\omega_r - \ell\omega_e^-(r)], \tag{2.10.41}$$

to evaluate

$$D_r(\omega_r) + iD_i(\omega_r) = \lim_{\delta \to 0_+} D(\omega_r + i\delta).$$

[Keep in mind that $D(\omega)$ is defined in Eq. (2.10.9) for Im $\omega > 0$.]

Since the spatial gradients are gentle in the vicinity of the density bump, both ω_i and $D_i(\omega_r)$ are small, and the term $\omega_i \, \partial D_i(\omega_r)/\partial \omega_r$ can be neglected in comparison with $D_r(\omega_r)$ in Eq. (2.10.38). Setting the real and imaginary parts of Eq. (2.10.38) equal to zero then gives

$$D_r(\omega_r) = 0 \tag{2.10.42}$$

and

$$\omega_i = -\frac{D_i(\omega_r)}{\partial D_r(\omega_r)/\partial \omega_r}. \tag{2.10.43}$$

Equation (2.10.42) can be interpreted as the dispersion relation that determines the oscillation frequency ω_r, assuming that the eigenfunctions $\delta\phi^\ell$ are known [see Eq. (2.10.39)]. Equation (2.10.43) determines the growth rate ω_i. Carrying out the r integration in Eq. (2.10.40), it is straightforward to show that Eq. (2.10.43) can be expressed as ($\ell \geqslant 1$ has been assumed)

$$\omega_i = -\frac{\pi}{\ell} \frac{\left[\dfrac{|\delta\phi^\ell|^2}{|\partial\omega_e^-(r)/\partial r|} \dfrac{\partial}{\partial r}\omega_{pe}^2(r) \right]_{r=r_s}}{P \displaystyle\int_0^{R_c} dr \, \frac{|\delta\phi^\ell|^2}{[\omega_r - \ell\omega_e^-(r)]^2} \frac{\partial}{\partial r}\omega_{pe}^2(r)}, \tag{2.10.44}$$

where ω_r is the solution to Eq. (2.10.42), and the resonant radius r_s is determined from

$$0 = \omega_r - \ell\omega_e^-(r_s). \tag{2.10.45}$$

It is assumed that r_s falls in the vicinity of the density bump in Fig. 2.10.3. Note that Eq. (2.10.45) corresponds to a resonance between the mode (ω_r, ℓ) and the equilibrium ($\mathbf{E}^0 \times \mathbf{B}_0$) drift motion of electrons at $r = r_s$. For the density profile illustrated in Fig. 2.10.3, the denominator in Eq. (2.10.44) is negative. Therefore the expression for the growth rate given in Eq. (2.10.44) corresponds to instability ($\omega_i > 0$) or stability ($\omega_i < 0$) accordingly as

$[\partial\omega_{pe}^2(r)/\partial r]_{r=r_s} > 0$ or $[\partial\omega_{pe}^2(r)/\partial r]_{r=r_s} < 0$, respectively. The similarity in structure between Eq. (2.10.44) and the Landau damping increment[135] for long-wavelength electron plasma osciallations is striking. Here $[\partial\omega_{pe}^2(r)/\partial r]_{r=r_s}$ plays the role of $[\partial F_{e0}(u)/\partial u]_{u=\omega_{pe}/|k|}$, that is, the instability is driven by configuration-space gradients rather than velocity-space gradients.

It is evident from Eqs. (2.10.39), (2.10.42), and (2.10.44) that the eigenfunctions $\delta\phi^\ell$ are needed in order to evaluate ω_r and ω_i explicitly. In general, this requires a detailed investigation of Poisson's equation for the perturbed potential, Eq. (2.10.7). However, for the gentle-bump configuration in Fig. 2.10.3, only a small error is incurred in calculating ω_r and ω_i if $\delta\phi^\ell$ is taken to correspond to the eigenfunctions *in the absence of the bump*, that is,

$$\delta\phi^\ell = Br^\ell, \qquad 0 < r < R_p, \tag{2.10.46}$$

$$\delta\phi^\ell = BR_p^{2\ell}\,\frac{R_c^{2\ell} - r^{2\ell}}{R_c^{2\ell} - R_p^{2\ell}}\,\frac{1}{r^\ell}, \quad R_p < r < R_c. \tag{2.10.47}$$

Equations (2.10.46) and (2.10.47), which follow from Eqs. (2.10.19) and (2.10.21) with $C = 0$, are the eigenfunctions for a solid beam ($R_0 = 0$). Approximating $\partial\omega_{pe}^2(r)/\partial r = -\overline{\omega}_{pe}^2\,\delta(r - R_p)$ in the integrand in Eq. (2.10.39), and making use of Eqs. (2.10.46) and (2.10.47), it is straightforward to show that the solution to $D(\omega_r) = 0$ [Eq. (2.10.42] is $\omega_r = \omega_{\rm D}[\ell - 1 + (R_p/R_c)^{2\ell}]$, which is identical to Eq. (2.10.37), as would be expected. Approximating $\partial\omega_{pe}^2(r)/\partial r = -\overline{\omega}_{pe}^2\,\delta(r - R_p)$ in the denominator in Eq. (2.10.44), we can express the growth rate ω_i as

$$\omega_i = \frac{\pi}{\ell}\,\frac{\left[\dfrac{|\delta\phi^\ell|^2}{|\partial\omega_e^-(r)/\partial r|}\,\dfrac{\partial}{\partial r}\omega_{pe}^2(r)\right]_{r=r_s}}{\overline{\omega}_{pe}^2\left[\dfrac{|\delta\phi^\ell|^2_{r=R_p}}{[\omega_r - \ell\omega_e^-(R_p)]^2}\right]}. \tag{2.10.48}$$

To sufficient accuracy for present purposes, $\omega_e^-(r)$ can be evaluated in Eq. (2.10.48) in the absence of a density bump. Making use of Eq. (2.10.6), we can express the resulting expression for $\omega_e^-(r)$ in the region $r \geqslant R_p$ as

$$\omega_e^-(r) = \omega_{\rm D}\,\frac{R_p^2}{r^2}, \quad r \geqslant R_p, \tag{2.10.49}$$

where $\omega_D \equiv \overline{\omega}_{pe}^2/2\Omega_e$. Substituting Eqs. (2.10.37) and (2.10.49) into the resonance condition, Eq. (2.10.45), gives

$$r_s^2 = \frac{\ell R_p^2}{\ell - 1 + (R_p/R_c)^{2\ell}} \cdot \qquad (2.10.50)$$

Note from Eq. (2.10.50) that $r_s > R_p$ for $\ell \geqslant 1$. When Eqs. (2.10.46), (2.10.47), (2.10.49), and (2.10.50) are substituted into Eq. (2.10.48), the growth rate ω_i reduces to[110]

$$\omega_i = \omega_D \frac{\pi}{2\ell}\left(\frac{R_p}{r_s}\right)^{2\ell-3}\left[1-\left(\frac{r_s}{R_c}\right)^{2\ell}\right]^2 \frac{R_p}{\overline{\omega}_{pe}^2}\left[\frac{\partial}{\partial r}\omega_{pe}^2(r)\right]_{r=r_s}, \qquad (2.10.51)$$

correct to lowest order. As illustrated in Fig. 2.10.3, if r_s is located in the region of positive slope, $[\partial\omega_{pe}^2(r)/\partial r]_{r=r_s} > 0$, then $\omega_i > 0$, which corresponds to instability. However, if r_s is located in the region of negative slope, $[\partial\omega_{pe}^2(r)/\partial r]_{r=r_s} < 0$, then $\omega_i < 0$ and the wave perturbation damps. Note from Eq. (2.10.51) that $|\omega_i| \ll \omega_D$ for small values of the density gradient.

2.11 INTERACTION OF RELATIVISTIC ELECTRON BEAMS WITH PLASMA

2.11.1 Equilibrium Configuration and Electrostatic Dispersion Relation

In this section the electrostatic dispersion relation, Eq. (2.7.15), is generalized to allow for relativistic axial motions of the plasma components along the external magnetic field $B_0\hat{\mathbf{e}}_z$. The following simplifying assumptions are made regarding the equilibrium configuration [see Section 2.1]. First, it is assumed that the equilibrium axial velocity profiles are independent of distance r from the axis of symmetry of the plasma column, that is,

$$c\beta_{\alpha z} = V_{\alpha z}^0 = \text{const. (independent of } r), \qquad (2.11.1)$$

for each plasma component. Second, as illustrated in Figure 2.11.1, it is assumed that the equilibrium density of each plasma component is constant out to some radius $r = R_p$ and is zero beyond, that is,

$$n_\alpha^0(r) = \begin{cases} \bar{n}_\alpha = \text{const.}, & 0 < r < R_p, \\ 0, & R_p < r < R_c, \end{cases} \tag{2.11.2}$$

where R_c denotes the radial location of a perfectly conducting wall. It is further assumed that the plasma is *electrically neutral* and that *no net axial current* is carried by the plasma components, that is,

$$\sum_\alpha \bar{n}_\alpha e_\alpha = 0, \tag{2.11.3}$$

and

$$\sum_\alpha \bar{n}_\alpha e_\alpha \beta_{\alpha z} c = 0. \tag{2.11.4}$$

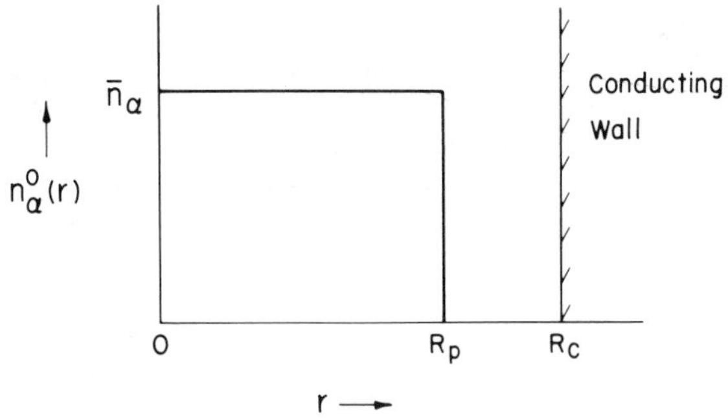

Fig. 2.11.1 Plot of $n_\alpha^0(r)$ versus r for the constant-density profile in Eq. (2.11.2). A grounded conducting wall is located at radius $r = R_c \geqslant R_p$.

It follows from Eqs. (2.1.3) and (2.11.3) that the equilibrium radial electric field is equal to zero,

$$E_r^0(r) = 0. \tag{2.11.5}$$

Moreover, from Eqs. (2.1.8) and (2.11.4) it follows that the equilibrium azimuthal self magnetic field is equal to zero,

$$B_\theta^s(r) = 0. \tag{2.11.6}$$

Physically, if one of the plasma components is a relativistic electron beam with axial current density $-e\bar{n}_b\beta_{bz}c$, Eq. (2.11.4) corresponds to the assumption that the background plasma carries a net return current $+e\bar{n}_b\beta_{bz}c$ that *magnetically neutralizes* the azimuthal self magnetic field produced by the relativistic electron beam [Eq. (2.11.6)]. Finally, it is assumed that there is no equilibrium azimuthal motion of the plasma components,

$$V^0_{\alpha\theta}(r) = r\omega_\alpha(r) = 0. \tag{2.11.7}$$

Therefore the equilibrium axial diamagnetic field $B^s_z(r)$ is equal to zero [see Eq. (2.1.7)],

$$B^s_z(r) = 0. \tag{2.11.8}$$

Note from Eqs. (2.11.5)-(2.11.8) that the equilibrium radial force equation (2.1.10) is trivially satisfied.

For the simple equilibrium configuration described by Eqs. (2.11.1)-(2.11.8), it is straightforward to extend the electrostatic stability analysis in Sections 2.7 and 2.8 to allow for relativistic axial motions of the plasma components along the external magnetic field $B_0\hat{\mathbf{e}}_z$. For present purposes it is adequate to quote the resulting dispersion relation[59] without presenting details of the derivation. Defining

$$\gamma_\alpha \equiv (1 - \beta^2_{\alpha z})^{-1/2}, \tag{2.11.9}$$

and making use of the fact that the plasma components have no equilibrium azimuthal motion ($\omega_\alpha = 0$), we obtain for Eq. (2.7.15) the modified form†

$$f_\ell - \left[1 - \sum_\alpha \frac{\bar{\omega}^2_{p\alpha}\gamma^{-1}_\alpha}{(\omega - k_z\beta_{\alpha z}c)^2 - \Omega^2_\alpha\gamma^{-2}_\alpha}\right] TR_p \frac{J'_\ell(TR_p)}{J_\ell(TR_p)}$$
$$= \ell \sum_\alpha \frac{\bar{\omega}^2_{p\alpha}\gamma^{-2}_\alpha\epsilon_\alpha\Omega_\alpha}{(\omega - k_z\beta_{\alpha z}c)\left[(\omega - k_z\beta_{\alpha z}c)^2 - \Omega^2_\alpha\gamma^{-2}_\alpha\right]} \tag{2.11.10}$$

where

$$f_\ell \equiv k_zR_p \frac{K_\ell(k_zR_c)I'_\ell(k_zR_p) - K'_\ell(k_zR_p)I_\ell(k_zR_c)}{K_\ell(k_zR_c)I_\ell(k_zR_p) - K_\ell(k_zR_p)I_\ell(k_zR_c)}, \tag{2.11.11}$$

and

---

†The factors $\bar{\omega}^2_{p\alpha}\gamma^{-1}_\alpha$ that occur in Eqs. (2.11.10) and (2.11.12) are incorrectly given as $\bar{\omega}^2_{p\alpha}\gamma^{-3}_\alpha$ in Ref. 59 (C. D. Striffler, private communication).

$$T^2 \equiv -k_z^2 \frac{\left[1 - \sum_\alpha \dfrac{\overline{\omega}_{p\alpha}^2 \gamma_\alpha^{-3}}{(\omega - k_z\beta_{\alpha z}c)^2}\right]}{\left[1 - \sum_\alpha \dfrac{\overline{\omega}_{p\alpha}^2 \gamma_\alpha^{-1}}{(\omega - k_z\beta_{\alpha z}c)^2 - \Omega_\alpha^2\gamma_\alpha^{-2}}\right]}. \tag{2.11.12}$$

In Eqs. (2.11.10)-(2.11.12), $\overline{\omega}_{p\alpha}^2 \equiv 4\pi\bar{n}_\alpha e_\alpha^2/m_\alpha$, $\Omega_\alpha \equiv |e_\alpha| B_0/m_\alpha c$, $\epsilon_\alpha \equiv \text{sgn}\, e_\alpha$, and the "prime" notation denotes derivatives with respect to the complete argument of the Bessel function, for example, $I_\ell'(k_z R_p) = [dI_\ell(x)/dx]_{x=k_z R_p}$. If $\beta_{\alpha z}^2 \ll 1$, then $\gamma_\alpha \simeq 1$, and Eq. (2.11.10) is identical to Eq. (2.7.15) with $\omega_\alpha = 0$, as expected. If there are relative axial drifts between plasma components, the dispersion relation, Eq. (2.11.10), supports unstable solutions with Im $\omega > 0$, at least in certain parameter regimes. For example, if one of the plasma components is a relativistic electron beam, the electron-electron two-stream instability may result from the relative drift motion between the beam electrons and the background plasma electrons.

No attempt will be made here to present a complete catalog of the two-stream instabilities associated with Eq. (2.11.10) in the different parameter regimes of interest. Rather, it is sufficient for present purposes to consider two limiting cases that illustrate the influence of finite radial geometry on two-stream instabilities resulting from relativistic beam-plasma interaction. The limiting cases considered are (*a*) the plasma-filled waveguide ($R_p = R_c$), and (*b*) the thin beam limit ($R_p \ll R_c$).

2.11.2 Two-Stream Instability in Plasma-Filled Waveguide

In this section the dispersion relation, Eq. (2.11.10), is examined in the limit where the plasma fills the waveguide,[59] that is, $R_p = R_c$. Taking the limit $R_p \to R_c$ in Eq. (2.11.10)† and paralleling the analysis in Section 2.8, it is straightforward to show that the dispersion relation reduces to

$$J_\ell(TR_c) = 0, \tag{2.11.13}$$

where T is defined in Eq. (2.11.12). It follows from Eq. (2.11.13) that

$$T^2 R_c^2 = p_{\ell m}^2, \tag{2.11.14}$$

†Note from Eq. (2.11.11) that $f_\ell \to \infty$ when $R_p \to R_c$.

where $p_{\ell m}$ is the mth zero of $J_\ell(x) = 0$. Substituting Eq. (2.11.12) into Eq. (2.11.14) and rearranging terms gives the dispersion relation,

$$0 = 1 - \frac{k_\perp^2}{k^2} \sum_\alpha \frac{\overline{\omega}_{p\alpha}^2 \gamma_\alpha^{-1}}{(\omega - k_z \beta_{\alpha z} c)^2 - \Omega_\alpha^2 \gamma_\alpha^{-2}}$$

$$- \frac{k_z^2}{k^2} \sum_\alpha \frac{\overline{\omega}_{p\alpha}^2 \gamma_\alpha^{-3}}{(\omega - k_z \beta_{\alpha z} c)^2}, \qquad (2.11.15)$$

where $k^2 \equiv k_z^2 + k_\perp^2$, and

$$k_\perp^2 \equiv \frac{p_{\ell m}^2}{R_c^2}. \qquad (2.11.16)$$

As an example of two-stream instability resulting from relativistic beam-plasma interaction, consider a three-component plasma that consists of beam electrons, background plasma electrons, and background plasma ions. The following notation is adopted throughout the remainder of Section 2.11.2:

$\alpha = b$ denotes beam electrons,

$\alpha = e$ denotes background plasma electrons,

$\alpha = i$ denotes background plasma ions.

Keep in mind that $\overline{n}_\alpha$ and $\beta_{\alpha z}$ are related by the charge neutrality condition and the current neutrality condition assumed in Eqs. (2.11.3) and (2.11.4). The dispersion relation, Eq. (2.11.15), simplifies in various limiting cases. For present purposes it is instructive to examine Eq. (2.11.15) in the frequency range

$$|\omega - k_z \beta_{\alpha z} c|^2 \ll \Omega_\alpha^2 \gamma_\alpha^{-2}, \quad \alpha = e, b, \qquad (2.11.17)$$

and

$$|\omega - k_z \beta_{\alpha z} c|^2 \gg \Omega_\alpha^2 \gamma_\alpha^{-2}, \quad \alpha = i. \qquad (2.11.18)$$

Equations (2.11.17) and (2.11.18) are statements that the electrons are strongly magnetized and the ions are weakly magnetized in the frequency range under investigation. It is further assumed that the magnetic field strength is sufficiently strong and/or the electron density is sufficiently low that

$$\Omega_\alpha^2 \gg \overline{\omega}_{p\alpha}^2 \gamma_\alpha, \; \alpha = e, b. \qquad (2.11.19)$$

Making use of Eqs. (2.11.17)-(2.11.19), and assuming that the ion motion is nonrelativistic ($\gamma_i \simeq 1$), we can express the dispersion relation, Eq. (2.11.15), in the approximate form

$$0 = 1 - \frac{\bar{\omega}_{pi}^2}{(\omega - k_z \beta_{iz} c)^2} - \frac{k_z^2}{k^2} \sum_{\alpha = e, b} \frac{\bar{\omega}_{p\alpha}^2 \gamma_\alpha^{-3}}{(\omega - k_z \beta_{\alpha z} c)^2} \cdot \tag{2.11.20}$$

Depending on the parameter regime, Eq. (2.11.20) can support unstable solutions with Im $\omega > 0$.

For purposes of illustration, we consider the electron-electron two-stream instability predicted by the dispersion relation, Eq. (2.11.20). For high frequencies well removed from resonance with the ions,

$$|\omega - k_z \beta_{iz} c|^2 \gg \bar{\omega}_{pi}^2, \tag{2.11.21}$$

Eq. (2.11.20) can be approximated by

$$0 = 1 - \frac{k_z^2}{k^2} \sum_{\alpha = e, b} \frac{\bar{\omega}_{p\alpha}^2 \gamma_\alpha^{-3}}{(\omega - k_z \beta_{\alpha z} c)^2} . \tag{2.11.22}$$

For appropriate values of parameters, Eq. (2.11.22) has one unstable solution with Im $\omega > 0$, which corresponds to an electron-electron two-stream instability produced by the relative drift motion between beam electrons and background plasma electrons. Using standard techniques, it is straightforward to show from Eq. (2.11.22) that instability exists only for axial wave vectors k_z that satisfy[141]

$$k_z^2 (\beta_{bz} - \beta_{ez})^2 c^2 < \bar{\omega}_{pb}^2 \gamma_b^{-3} \frac{k_z^2}{k^2} \left[1 + \left(\frac{\bar{n}_e}{\bar{n}_b} \right)^{1/3} \frac{\gamma_b}{\gamma_e} \right]^3 , \tag{2.11.23}$$

where $k^2 \equiv k_z^2 + k_\perp^2$ [see Eq. (2.11.16)], $\alpha = b$ refers to beam electrons, and $\alpha = e$ refers to background plasma electrons. Since $k_\perp^2 = p_{\ell m}^2 / R_c^2$, Eq. (2.11.23) can be expressed in the equivalent form

$$k_z^2 (\beta_{bz} - \beta_{ez})^2 c^2 < \bar{\omega}_{pb}^2 \gamma_b^{-3} \left[1 + \left(\frac{\bar{n}_e}{\bar{n}_b} \right)^{1/3} \frac{\gamma_b}{\gamma_e} \right]^3 - \frac{p_{\ell m}^2}{R_c^2} (\beta_{bz} - \beta_{ez})^2 c^2 . \tag{2.11.24}$$

Since k_z is assumed to be real, a necessary condition for instability is that the right-hand side of Eq. (2.11.24) be positive, that is,

$$\bar{\omega}_{pb}^2 > \frac{(p_{\ell m} / R_c)^2 \gamma_b^3 (\beta_{bz} - \beta_{ez})^2 c^2}{\left[1 + \left(\frac{\bar{n}_e}{\bar{n}_b}\right)^{1/3} \frac{\gamma_b}{\gamma_e}\right]^3}. \tag{2.11.25}$$

For fixed values of $\bar{n}_b$, $\bar{n}_e$, γ_b, and γ_e, the instability condition, Eq. (2.11.25), cannot be satisfied if R_c is sufficiently small, that is, finite radial geometry effects have a stabilizing influence on the two-stream instability. In terms of the magnitude of the beam current, $I_b \equiv |\bar{n}_b e \beta_{bz} c| \pi R_p^2$, it is straightforward to show that the instability condition, Eq. (2.11.25), can be expressed in the equivalent form,

$$I_b > I_{\text{crit}}, \tag{2.11.26}$$

where

$$I_{\text{crit}} = I_A \frac{p_{\ell m}^2 \beta_{bz}^2 \gamma_b^2 (1 - \beta_{ez} / \beta_{bz})^2}{4\left[1 + \left(\frac{\bar{n}_e}{\bar{n}_b}\right)^{1/3} \frac{\gamma_b}{\gamma_e}\right]^3}. \tag{2.11.27}$$

In Eq. (2.11.27), the critical current I_{crit} has been expressed in units of the Alfvén critical current,[142] $I_A = |(m_e c^3 / e) \beta_{bz} \gamma_b| \simeq 17{,}000 |\beta_{bz}| \gamma_b$ amperes. Note from Eq. (2.11.27) that the critical beam current for instability can be substantial. For example, in the case of symmetric equidensity electron components, $\bar{n}_e = \bar{n}_b$ and $\beta_{ez} = -\beta_{bz}$, the instability condition, Eq. (2.11.26), reduces to

$$I_b > I_A \frac{p_{\ell m}^2 \gamma_b^2 \beta_{bz}^2}{8}. \tag{2.11.28}$$

The lowest current threshold is for $\ell = 0$ and $m = 1$. Substituting $p_{01} \simeq 2.4$ into Eq. (2.11.28) gives

$$I_b > 0.72 \gamma_b^2 \beta_{bz}^2 I_A. \tag{2.11.29}$$

It follows from Eq. (2.11.29) that the critical beam current for instability exceeds the Alfvén current I_A whenever $|\beta_{bz}| \gtrsim 0.76$.

In general, the exact solution to Eq. (2.11.22) for the complex oscillation frequency $\omega = \omega_r + i\omega_i$ is not tractable. However, for the special case of symmetric equidensity electron components with $\bar{n}_e = \bar{n}_b$ and $\beta_{ez} = -\beta_{bz}$, closed expressions for ω_r and ω_i can be obtained. When $\bar{n}_e = \bar{n}_b$ and $\beta_{ez} = -\beta_{bz}$ are substituted into Eq. (2.11.22), it is straightforward to show that the solution for the unstable branch with $\omega_i = \text{Im}\,\omega > 0$ is given by $\omega_r = 0$ and

$$\omega_i = |k_z\beta_{bz}c| \left(\left\{ \left[1 + \frac{\overline{\omega}_{pb}^2\gamma_b^{-3}}{(k_z^2 + k_\perp^2)\beta_{bz}^2 c^2} \right]^2 - \left[1 - \frac{2\overline{\omega}_{pb}^2\gamma_b^{-3}}{(k_z^2 + k_\perp^2)\beta_{bz}^2 c^2} \right] \right\}^{1/2} - \left[1 + \frac{\overline{\omega}_{pb}^2\gamma_b^{-3}}{(k_z^2 + k_\perp^2)\beta_{bz}^2 c^2} \right] \right)^{1/2}, \tag{2.11.30}$$

when the threshold condition in Eq. (2.11.28) is satisfied. In the limit where the beam radius is sufficiently large that the inequality $k_\perp^2\beta_{bz}^2c^2 \ll 8\overline{\omega}_{pb}^2\gamma_b^{-3}$ is satisfied, it is straightforward to show that the maximum growth rate obtained from Eq. (2.11.30) can be approximated by

$$[\omega_i]_{\max} = \frac{1}{2}\overline{\omega}_{pb}\gamma_b^{-3/2}, \tag{2.11.31}$$

and the corresponding axial wave vector at maximum growth is

$$[k_z^2]_{\max} = \frac{3}{4}\frac{\overline{\omega}_{pb}^2\gamma_b^{-3}}{\beta_{bz}^2c^2}. \tag{2.11.32}$$

In deriving the dispersion relation, Eq. (2.11.15), it was assumed that $R_p = R_c$. In concluding this section, it should be noted that Eq. (2.11.15) is also approximately correct when $R_p \neq R_c$ *provided* the vacuum gap between the conducting wall and the surface of the plasma column is sufficiently small, that is, $(R_c - R_p)/R_c \ll 1$. A careful examination of Eq. (2.11.10) shows that Eq. (2.11.15) is a valid approximation for the electrostatic dispersion relation as long as

$$\frac{f_\ell}{p_{\ell m}} \gg 1. \tag{2.11.33}$$

For example, for long wavelength perturbations with $k_z^2R_c^2 \ll 1, f_\ell$ can be approximated by

$$f_\ell = \begin{cases} \dfrac{1}{\ell n\,(R_c/R_p)}, & \ell = 0, \\[2ex] \ell\,\dfrac{1 + (R_p/R_c)^{2\ell}}{1 - (R_p/R_c)^{2\ell}}, & \ell \neq 0. \end{cases} \tag{2.11.34}$$

Therefore, for $k_z^2 R_c^2 \ll 1$, Eq. (2.11.33) can be expressed as

$$p_{0m}\ \ell\text{n}(R_c / R_p) \ll 1, \quad \text{for } \ell = 0, \tag{2.11.35}$$

and

$$1 - (R_p / R_c)^{2\ell} \ll \frac{2\ell}{p_{\ell m} + \ell}, \quad \text{for } \ell \neq 0. \tag{2.11.36}$$

For $\ell = 0$ and $m = 1$, Eq. (2.11.35) gives the condition

$$p_{01}\,\ell\text{n}\,(R_c / R_p) \simeq 2.4\ \ell\text{n}\,(R_c / R_p) \ll 1$$

for validity of Eq. (2.11.15).

2.11.3 Two-Stream Instability in Thin Beam Limit

In this section the dispersion relation, Eq. (2.11.10), is examined in the limit where the radius of the plasma column is much smaller than the radius of the waveguide,[59]

$$R_p \ll R_c. \tag{2.11.37}$$

It is also assumed that the axial wavelength of the perturbation is sufficiently long that

$$k_z^2 R_c^2 \ll 1, \tag{2.11.38}$$

and

$$R_p^2 |T|^2 \ll 1, \tag{2.11.39}$$

where T is defined in Eq. (2.11.12). Making use of Eqs. (2.11.37) and (2.11.38), it is straightforward to show that f_ℓ, defined in Eq. (2.11.11), can be approximated by

$$f_\ell = \begin{cases} \dfrac{1}{\ell\text{n}\,(R_c/R_p)}, & \ell = 0, \\[2ex] \ell, & \ell \neq 0. \end{cases} \tag{2.11.40}$$

Furthermore, for $|T|^2 R_p^2 \ll 1$, it is valid to approximate

$$\left[x \frac{J'_\ell(x)}{J_\ell(x)}\right]_{x=TR_p} = \ell \left[1 - \frac{T^2 R_p^2}{2\ell(\ell+1)} + \ldots\right] \qquad (2.11.41)$$

in the dispersion relation, Eq. (2.11.10). When (2.11.41) is substituted into Eq. (2.11.10), the dispersion relation can be expressed in the approximate form

$$f_\ell - \ell \left[1 - \sum_\alpha \frac{\overline{\omega}_{p\alpha}^2 \gamma_\alpha^{-1}}{(\omega - k_z\beta_{\alpha z}c)^2 - \Omega_\alpha^2\gamma_\alpha^{-2}}\right]$$

$$+ \frac{k_z^2 R_p^2}{2(\ell+1)} \left[1 - \sum_\alpha \frac{\overline{\omega}_{p\alpha}^2 \gamma_\alpha^{-3}}{(\omega - k_z\beta_{\alpha z}c)^2}\right]$$

$$= \ell \sum_\alpha \frac{\overline{\omega}_{p\alpha}^2 \gamma_\alpha^{-2} \epsilon_\alpha \Omega_\alpha}{(\omega - k_z\beta_{\alpha z}c)\,[(\omega - k_z\beta_{\alpha z}c)^2 - \Omega_\alpha^2\gamma_\alpha^{-2}]}, \qquad (2.11.42)$$

where f_ℓ is defined in Eq. (2.11.40).

For appropriate values of parameters, Eq. (2.11.42) supports unstable solutions with Im $\omega > 0$. Instability can exist for both $\ell = 0$ and $\ell \neq 0$. Note that the right-hand side of Eq. (2.11.42), which results from an accumulation of perturbed surface charge at $r = R_p$, makes a nonzero contribution to the dispersion relation when $\ell \neq 0$. Rather than present a comprehensive analysis of Eq. (2.11.42), it is sufficient for present purposes to examine only the case of azimuthally symmetric perturbations with $\ell = 0$. The reader is referred to the extensive stability analyses in the literature which investigate Eq. (2.11.42) for the case $\ell \neq 0$ (see, for example, Reference 59). Making use of Eq. (2.11.40), we reduce the dispersion relation to

$$\ln(R_c/R_p) + \frac{k_z^2 R_p^2}{2}\left[1 - \sum_\alpha \frac{\overline{\omega}_{p\alpha}^2 \gamma_\alpha^{-3}}{(\omega - k_z\beta_{\alpha z}c)^2}\right] = 0 \qquad (2.11.43)$$

for $\ell = 0$. Defining

$$k_\perp^2 \equiv (2/R_p^2)\ln(R_c/R_p), \qquad (2.11.44)$$

we can express Eq. (2.11.43) as

$$0 = 1 - \frac{k_z^2}{k^2}\sum_\alpha \frac{\overline{\omega}_{p\alpha}^2 \gamma_\alpha^{-3}}{(\omega - k_z\beta_{\alpha z}c)^2} \qquad (2.11.45)$$

where $k^2 \equiv k_z^2 + k_\perp^2$. Formally, Eq. (2.11.45) is identical in structure to Eq. (2.11.20) with a redefinition of $k_\perp^2$ [compare Eqs. (2.11.16) and (2.11.44)].

If it is assumed that the plasma consists of beam electrons ($\alpha = b$), background plasma ions ($\alpha = i$), and background plasma electrons ($\alpha = e$), the analysis of the electron-electron two-stream instability proceeds in the same manner as outlined in Section 2.11.2 [see the discussion that follows Eq. (2.11.20)]. The essential modification is that $k_\perp^2 = p_{\ell m}^2/R_c^2$ is replaced by $k_\perp^2 = (2/R_p^2) \times \ell n\,(R_c/R_p)$. To illustrate two-stream instability produced by the relative drift motion between electrons and ions, it is instructive to examine the case in which *all* of the electrons are drifting with axial velocity $\beta_{bz}c$. When $\bar{n}_e = 0$ is substituted into Eq. (2.11.45), the dispersion relation for $\ell = 0$ becomes

$$0 = 1 - \frac{k_z^2}{k^2}\frac{\bar{\omega}_{pi}^2\gamma_i^{-3}}{(\omega - k_z\beta_{ic}c)^2} - \frac{k_z^2}{k^2}\frac{\bar{\omega}_{pb}^2\gamma_b^{-3}}{(\omega - k_z\beta_{bz}c)^2}. \tag{2.11.46}$$

For appropriate values of parameters, Eq. (2.11.46) has one unstable solution with Im $\omega > 0$. Using standard techniques,[141] it is straightforward to show from Eq. (2.11.46) that instability exists only for axial wave vectors k_z that satisfy [see also Eq. (2.11.24)]

$$k_z^2(\beta_{bz} - \beta_{iz})^2c^2 < \bar{\omega}_{pb}^2\gamma_b^{-3}\left[1 + \left(\frac{Zm_e}{m_i}\right)^{1/3}\frac{\gamma_b}{\gamma_i}\right]^3 - \frac{2}{R_p^2\,\ell n\,(R_c/R_p)}(\beta_{bz} - \beta_{iz})^2c^2\,, \tag{2.11.47}$$

where $\bar{n}_b = Z\bar{n}_i$ (Z = degree of ionization) has been assumed. Since k_z is assumed to be real, a necessary condition for instability is that the right-hand side of Eq. (2.11.47) be positive. In terms of the magnitude of the beam current, $I_b \equiv |\bar{n}_b e\beta_{bz}c|\pi R_p^2$, it is straightforward to show that the necessary condition for instability can be expressed as

$$I_b > I_{\text{crit}} \equiv I_A\,\frac{1}{2\ell n\,(R_c/R_p)}\,\frac{\beta_{bz}^2\gamma_b^2(1 - \beta_{iz}/\beta_{bz})^2}{\left[1 + \left(\frac{Zm_e}{m_i}\right)^{1/3}\frac{\gamma_b}{\gamma_i}\right]^3}, \tag{2.11.48}$$

where $I_A \equiv |(m_ec^3/e)\beta_{bz}\gamma_b|$ is the Alfvén critical current. Note from Eq. (2.11.48) that an increase in R_c/R_p tends to *reduce* the current threshold for instability.

When $(Zm_e/m_i)^{1/3}\gamma_b/\gamma_i \ll 1$, and the beam radius is sufficiently large that the inequality $k_\perp^2(\beta_{bz}-\beta_{iz})^2c^2 \ll \bar{\omega}_{pb}^2\gamma_b^{-3}$ is satisfied, it is straightforward to show that the maximum growth rate obtained from Eq. (2.11.46) can be approximated by

$$[\omega_i]_{\max} = \frac{\sqrt{3}}{2}\left(\frac{Zm_e}{2m_i}\right)^{1/3}\frac{\gamma_b}{\gamma_i}\bar{\omega}_{pb}\gamma_b^{-3/2}. \tag{2.11.49}$$

Furthermore, the corresponding axial wave vector and phase velocity at maximum growth are approximately

$$[k_z^2]_{\max} = \frac{\bar{\omega}_{pb}^2\gamma_b^{-3}}{(\beta_{bz}-\beta_{iz})^2c^2} \tag{2.11.50}$$

and

$$\left[\frac{\omega_r}{k_z}-\beta_{iz}c\right]_{\max} = \frac{1}{2}\left(\frac{Zm_e}{2m_i}\right)^{1/3}\frac{\gamma_b}{\gamma_i}(\beta_{bz}-\beta_{iz})c. \tag{2.11.51}$$

Note that the results in Eqs. (2.11.49) – (2.11.51) are independent of beam radius for the limiting case considered here.

CHAPTER 3

VLASOV EQUILIBRIA AND STABILITY

3.1 VLASOV EQUILIBRIA FOR AXISYMMETRIC SYSTEMS

3.1.1 General Discussion

In Chapter 2, the equilibrium and stability properties of plasmas with equilibrium electric and/or magnetic self fields were examined within the framework of the macroscopic fluid equations. As elaborated in Section 1.3, the cold-fluid description gives valuable insight into equilibrium and stability properties that do not depend on the detailed momentum-space structure of the equilibrium distribution function $f_\alpha^0(\mathbf{x}, \mathbf{p})$. In order to include the effects of finite temperature, in this chapter the equilibrium and stability properties of collisionless plasmas with equilibrium electric and/or magnetic self fields are examined within the framework of the Vlasov-Maxwell equations. In this case, the one-particle distribution function, $f_\alpha(\mathbf{x}, \mathbf{p}, t)$, and the electric and magnetic fields, $\mathbf{E}(\mathbf{x}, t)$ and $\mathbf{B}(\mathbf{x}, t)$, evolve self-consistently according to Eq. (1.3.2) and Eqs. (1.3.4)-(1.3.7). In Sections 3.2–3.7 it is shown that the equilibrium and stability properties are greatly influenced by the choice of equilibrium distribution function $f_\alpha^0(\mathbf{x}, \mathbf{p})$. Moreover, there is a broad class of plasma-like waves and instabilities that depend on the $\mathbf{p}$-space structure of $f_\alpha^0(\mathbf{x}, \mathbf{p})$, for example, Landau damping (or growth) resulting from resonant wave-particle interaction, and wave dispersion produced by thermal effects (see Section 3.7). For a brief review of the fundamentals of the Vlasov-Maxwell formalism[87–92] used throughout Chapter 3, the reader is referred to Sections 1.3.1 and 1.3.2.

In this section, the general procedure for constructing self-consistent Vlasov equilibria from the steady-state ($\partial/\partial t = 0$) Vlasov-Maxwell equations is discussed for two types of *axisymmetric* equilibrium configurations, illustrated in Figs. 3.1.1 and 3.1.2.†

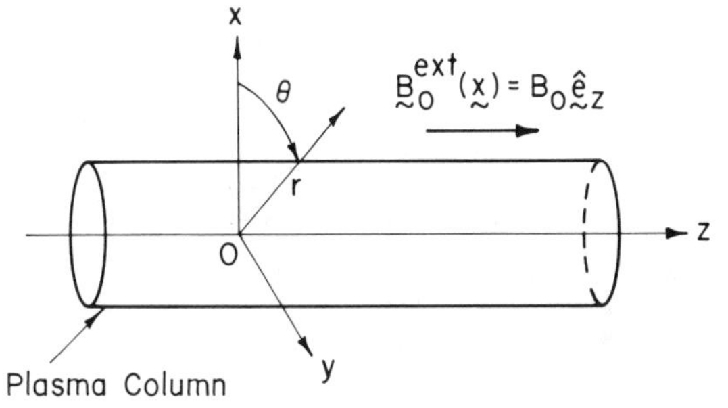

Fig. 3.1.1 Axisymmetric equilibrium configuration for a plasma column of infinite axial extent, aligned parallel to a uniform external magnetic field $\mathbf{B}_0^{\text{ext}}(\mathbf{x}) = B_0\hat{\mathbf{e}}_z$. Cylindrical polar coordinates (r, θ, z) are introduced with the z-axis coinciding with the axis of symmetry; θ is the polar angle in the x-y plane, and $r = \sqrt{x^2 + y^2}$ is the radial distance from the z-axis.

The equilibrium configuration illustrated in Fig. 3.1.1 corresponds to a plasma column of infinite axial extent aligned parallel to a uniform external magnetic field, $\mathbf{B}_0^{\text{ext}}(\mathbf{x}) = B_0\hat{\mathbf{e}}_z$. Equilibrium properties are assumed to be independent of $z(\partial/\partial z = 0)$ and azimuthally symmetric $(\partial/\partial\theta = 0)$ about an axis of symmetry parallel to $B_0\hat{\mathbf{e}}_z$. The equilibrium configuration illustrated in Fig. 3.1.1 may represent, for example, the following plasma systems:

1. Magnetically confined, nonneutral plasma column in which the particle motions are nonrelativistic (Section 3.2).
2. Relativistic E-layer in a neutralizing plasma background (Section 3.3).
3. Relativistic electron beam propagating through a partially neutralizing ion background, with and without a magnetic guide field (Section 3.4).

The equilibrium configuration illustrated in Fig. 3.1.2 corresponds to a plasma radially and axially confined in an external mirror field $\mathbf{B}_0^{\text{ext}}(\mathbf{x})$. As in Fig. 3.1.1, equilibrium properties are assumed to be azimuthally symmetric $(\partial/\partial\theta = 0)$ about the z-axis. The equilibrium configuration illustrated in Fig. 3.1.2 may represent, for example, the following plasma systems:

† Cylindrical polar coordinates (r, θ, z) are used throughout Chapter 3 (see Figs. 3.1.1 and 3.1.2). A vector $\mathbf{V}$ is represented as $\mathbf{V} = V_r\hat{\mathbf{e}}_r + V_\theta\hat{\mathbf{e}}_\theta + V_z\hat{\mathbf{e}}_z$, where $\hat{\mathbf{e}}_r$, $\hat{\mathbf{e}}_\theta$ and $\hat{\mathbf{e}}_z$ are unit vectors in the r-, θ-, and z-directions, respectively. Moreover, the gradient operator is expressed as $\nabla = \hat{\mathbf{e}}_r(\partial/\partial r) + \hat{\mathbf{e}}_\theta(1/r)(\partial/\partial\theta) + \hat{\mathbf{e}}_z(\partial/\partial z)$.

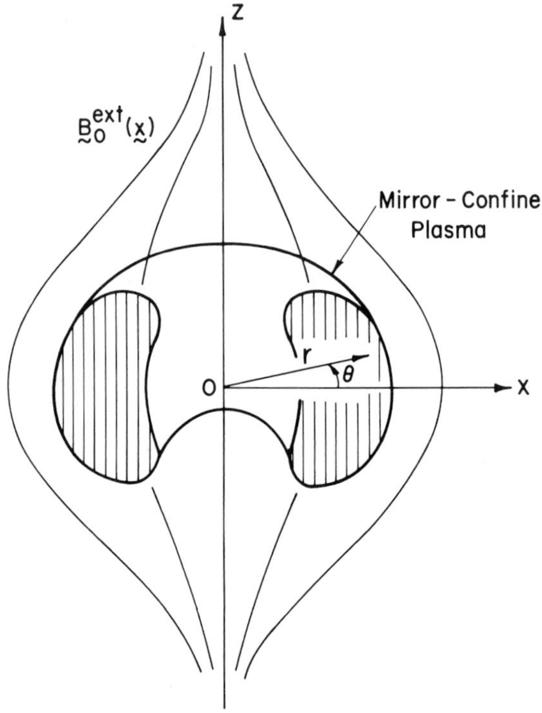

Fig. 3.1.2 Axisymmetric equilibrium configuration for a toroidal plasma radially and axially confined in an external mirror field $\mathbf{B}_0^{ext}(\mathbf{x})$. Cylindrical polar coordinates (r, θ, z) are introduced with the z-axis coinciding with the axis of symmetry, and $z = 0$ at the mirror midplane; r is the radial distance from the axis of symmetry, and θ is the polar angle.

1. Magnetically confined nonneutral plasma in which the particle motions are nonrelativistic.[75]
2. Magnetically confined, partially neutralized, relativistic electron ring (Section 3.5).

In Sections 3.1.2 and 3.1.3, the equilibrium equations are obtained for the axisymmetric equilibrium configurations illustrated in Figs. 3.1.1 and 3.1.2, respectively. Keep in mind that the single-particle constants of the motion in the equilibrium fields are central to an equilibrium theory based on the Vlasov-Maxwell equations (see Section 1.3.2). In particular, any distribution function $f_\alpha^0(\mathbf{x}, \mathbf{p})$ that is a function only of the single-particle constants of the motion in the equilibrium fields satisfies the steady-state $(\partial/\partial t = 0)$ Vlasov equation, Eq. (1.3.8).

3.1.2 Axisymmetric Column Equilibria

In this section the equilibrium equations are obtained for the axisymmetric equilibrium configuration illustrated in Fig. 3.1.1. In general, the plasma column is electrically nonneutral. Therefore there is an equilibrium electric field,

$\mathbf{E}^0(\mathbf{x}) = -\nabla\phi^0(\mathbf{x})$, produced self-consistently by the deviation from charge neutrality in equilibrium [see Eq. (1.3.11)]. Since the equilibrium is uniform in the z-direction ($\partial/\partial z = 0$) and azimuthally symmetric ($\partial/\partial\theta = 0$) about the z-axis, the equilibrium electric field can be expressed as

$$\mathbf{E}^0(\mathbf{x}) = E_r^0(r)\hat{\mathbf{e}}_r, \tag{3.1.1}$$

where r is the radial distance from the axis of symmetry, and $\hat{\mathbf{e}}_r$ is a unit vector in the r-direction. Since $\mathbf{E}^0(\mathbf{x}) = -\nabla\phi^0(\mathbf{x})$, the radial electric field can be expressed as

$$E_r^0(r) = -\frac{\partial}{\partial r}\phi^0(r), \tag{3.1.2}$$

where $\phi^0(r)$ is the electrostatic potential.

The plasma components making up the equilibrium configuration in Fig. 3.1.1 in general have mean motions in both the axial (z) and azimuthal (θ) directions. If the plasma carries an equilibrium axial current, $J_z^0(r)$ (e.g., due to the passage of a relativistic electron beam), there is a corresponding azimuthal self magnetic field, $B_\theta^s(r)$, generated self-consistently. Similarly, if the plasma carries an equilibrium azimuthal current, $J_\theta^0(r)$ [e.g., due to the rotation of plasma components produced by the radial electric field $E_r^0(r)$], there is a corresponding axial self magnetic field, $B_z^s(r)$, generated self-consistently. The total equilibrium magnetic field, $\mathbf{B}^0(\mathbf{x})$, can be expressed as

$$\mathbf{B}^0(\mathbf{x}) = B_0\hat{\mathbf{e}}_z + B_\theta^s(r)\hat{\mathbf{e}}_\theta + B_z^s(r)\hat{\mathbf{e}}_z, \tag{3.1.3}$$

where $\hat{\mathbf{e}}_z$ and $\hat{\mathbf{e}}_\theta$ are unit vectors in the z-direction and θ-direction, respectively. In terms of the vector potentials for the external and self magnetic fields, $\mathbf{B}_0^{\text{ext}}(\mathbf{x}) = B_0\hat{\mathbf{e}}_z$ and $\mathbf{B}_0^s(\mathbf{x}) = B_\theta^s(r)\hat{\mathbf{e}}_\theta + B_z^s(r)\hat{\mathbf{e}}_z$ can be represented as

$$\mathbf{B}_0^{\text{ext}}(\mathbf{x}) = \nabla\times\mathbf{A}_0^{\text{ext}}(\mathbf{x}); \quad \mathbf{B}_0^s(\mathbf{x}) = \nabla\times\mathbf{A}_0^s(\mathbf{x}), \tag{3.1.4}$$

where $\nabla = \hat{\mathbf{e}}_r(\partial/\partial r) + \hat{\mathbf{e}}_\theta(1/r)(\partial/\partial\theta)$, and

$$\mathbf{A}_0^{\text{ext}}(\mathbf{x}) = A_\theta^{\text{ext}}(r)\hat{\mathbf{e}}_\theta = \frac{rB_0}{2}\hat{\mathbf{e}}_\theta; \qquad \mathbf{A}_0^s(\mathbf{x}) = A_\theta^s(r)\hat{\mathbf{e}}_\theta + A_z^s(r)\hat{\mathbf{e}}_z. \tag{3.1.5}$$

From Eqs. (3.1.4) and (3.1.5) the components of the equilibrium self magnetic field can be expressed in terms of $A_\theta^s(r)$ and $A_z^s(r)$ as

$$B_z^s(r) = \frac{1}{r}\frac{\partial}{\partial r}rA_\theta^s(r), \tag{3.1.6}$$

$$B_\theta^s(r) = -\frac{\partial}{\partial r}A_z^s(r). \tag{3.1.7}$$

Central to an equilibrium theory based on the steady-state ($\partial/\partial t = 0$) Vlasov-Maxwell equations are the single-particle constants of the motion in the equilibrium fields.[137] For the equilibrium field configuration given in Eqs. (3.1.1) and (3.1.3), the single-particle constants of the motion are the total energy H, the canonical angular momentum P_θ, and the canonical momentum in the z-direction P_z, where

$$H \equiv (m_\alpha^2 c^4 + c^2 \mathbf{p}^2)^{1/2} + e_\alpha \phi^0(r), \tag{3.1.8}$$

$$P_\theta \equiv r\left[p_\theta + \frac{e_\alpha}{c} A_\theta^{\text{ext}}(r) + \frac{e_\alpha}{c} A_\theta^s(r)\right], \tag{3.1.9}$$

$$P_z \equiv p_z + \frac{e_\alpha}{c} A_z^s(r). \tag{3.1.10}$$

In Eqs. (3.1.8)-(3.1.10), e_α and m_α are the charge and rest mass, respectively, of a particle of species α, c is the speed of light *in vacuo*, $\mathbf{p}$ is the mechanical momentum, and $\mathbf{p}^2 \equiv p_r^2 + p_\theta^2 + p_z^2$.† Equations (3.1.8)-(3.1.10) are relativistically correct expressions for the single-particle constants of the motion. Moreover, the particle velocity $\mathbf{v}$ is related to the mechanical momentum $\mathbf{p}$ by

$$\mathbf{v} = \frac{\mathbf{p}/m_\alpha}{(1 + \mathbf{p}^2/m_\alpha^2 c^2)^{1/2}}. \tag{3.1.11}$$

In the spirit of an equilibrium theory based on the Vlasov-Maxwell equations, any distribution function $f_\alpha^0(\mathbf{x}, \mathbf{p})$ that depends only on the single-particle constants of the motion, H, P_θ, and P_z, is a solution to the steady-state ($\partial/\partial t = 0$) Vlasov equation, that is, equilibrium distribution functions of the form

$$f_\alpha^0(\mathbf{x}, \mathbf{p}) = f_\alpha^0(H, P_\theta, P_z) \tag{3.1.12}$$

are solutions to Eq (1.3.8). This result can be verified directly by substituting Eq. (3.1.12) into Eq. (1.3.8) and making use of Eqs. (3.1.1)-(3.1.10). Evidently there is considerable lattitude in the specific choice of functional form for $f_\alpha^0(H, P_\theta, P_z)$. Once $f_\alpha^0(H, P_\theta, P_z)$ is specified, however, the equilibrium electrostatic potential and self-field vector potential can be determined *self-consistently* from the steady-state Maxwell equations, Eqs. (1.3.10) and (1 3.11). Making use of Eqs. (1.3.10), (1.3.11), ‡and (3.1.1)-(3.1.7), it is straightforward to show that $\phi^0(r)$, $A_\theta^s(r)$, and $A_z^s(r)$ are determined self-consistently in terms of $f_\alpha^0(H, P_\theta, P_z)$ from

$$\frac{1}{r}\frac{\partial}{\partial r} r \frac{\partial}{\partial r} \phi^0(r) = -4\pi \sum_\alpha e_\alpha \int d^3p\, f_\alpha^0(H, P_\theta, P_z), \tag{3.1.13}$$

$$\frac{\partial}{\partial r}\frac{1}{r}\frac{\partial}{\partial r} r A_\theta^s(r) = -\frac{4\pi}{c} \sum_\alpha e_\alpha \int d^3p\, v_\theta f_\alpha^0(H, P_\theta, P_z), \tag{3.1.14}$$

$$\frac{1}{r}\frac{\partial}{\partial r} r \frac{\partial}{\partial r} A_z^s(r) = -\frac{4\pi}{c} \sum_\alpha e_\alpha \int d^3p\, v_z f_\alpha^0(H, P_\theta, P_z), \tag{3.1.15}$$

---

†Throughout Chapter 3, lower case $\mathbf{p}$ is used to denote mechanical momentum, whereas upper case P denotes the canonical angular momentum P_θ and the axial canonical momentum P_z.

---

‡Throughout Chapter 3, it is assumed that the external charge density is equal to zero, that is, $\rho_{\text{ext}}(\mathbf{x}) = 0$, in the equilibrium Poisson equation, Eq. (1.3.11).

where $v_j = (p_j/m_\alpha)(1 + \mathbf{p}^2/m_\alpha^2 c^2)^{-1/2}$ [see Eq. (3.1.11)]. It is important to note that Eqs. (3.1.13)-(3.1.15) are generally nonlinear differential equations for the equilibrium self-field potentials. This follows since H, P_θ, and P_z depend on $\phi^0(r)$, $A_\theta^s(r)$, and $A_z^s(r)$ [see Eqs. (3.1.8)-(3.1.10)].

For axisymmetric column equilibria (Fig. 3.1.1), the procedure for constructing self-consistent equilibria from the steady-state Vlasov-Maxwell equations can be summarized as follows. First, specify a functional form for $f_\alpha^0(H, P_\theta, P_z)$. Second, using this form for $f_\alpha^0(H, P_\theta, P_z)$, calculate the equilibrium charge density $\rho^0(r)$ and the equilibrium current densities $J_\theta^0(r)$ and $J_z^0(r)$, where†

$$\rho^0(r) = \sum_\alpha e_\alpha n_\alpha^0(r) = \sum_\alpha e_\alpha \int d^3p\, f_\alpha^0(H, P_\theta, P_z), \tag{3.1.16}$$

$$J_\theta^0(r) = \sum_\alpha e_\alpha n_\alpha^0(r) V_{\alpha\theta}^0(r) = \sum_\alpha e_\alpha \int d^3p\, v_\theta f_\alpha^0(H, P_\theta, P_z), \tag{3.1.17}$$

$$J_z^0(r) = \sum_\alpha e_\alpha n_\alpha^0(r) V_{\theta z}^0(r) = \sum_\alpha e_\alpha \int d^3p\, v_z f_\alpha^0(H, P_\theta, P_z). \tag{3.1.18}$$

Finally, making use of the expressions obtained for $\rho^0(r)$, $J_\theta^0(r)$, and $J_z^0(r)$, calculate the equilibrium self-field potentials self-consistently from Eqs. (3.1.13)-(3.1.15). Once $\phi^0(r)$, $A_\theta^s(r)$, and $A_z^s(r)$ are known, other equilibrium properties (e.g., the particle stress tensor) can be calculated explicitly from the equilibrium distribution function, $f_\alpha^0(H, P_\theta, P_z)$.

The equilibrium analysis simplifies considerably in situations where the particle motions are nonrelativistic[76–78] and the axial and azimuthal self magnetic fields, $B_z^s(r)$ and $B_\theta^s(r)$, are negligibly small (see Section 3.2). In this case the single-particle constants of the motion defined in Eqs. (3.1.8)-(3.1.10) can be approximated by

$$H = m_\alpha c^2 + \frac{\mathbf{p}^2}{2m_\alpha} + e_\alpha \phi^0(r), \tag{3.1.19}$$

$$P_\theta = r\left[p_\theta + \frac{e_\alpha}{c} A_\theta^{\text{ext}}(r)\right] = r\left[p_\theta + m_\alpha r \frac{\epsilon_\alpha \Omega_\alpha}{2}\right], \tag{3.1.20}$$

$$P_z = p_z, \tag{3.1.21}$$

where $\epsilon_\alpha \equiv \text{sgn}\, e_\alpha$, and $\Omega_\alpha \equiv |e_\alpha| B_0/m_\alpha c$. Note that $A_\theta^s(r)$ and $A_z^s(r)$, the components of vector potential for the equilibrium self magnetic field, have been neglected in Eqs. (3.1.20) and (3.1.21). Insofar as the equilibrium is electrically nonneutral and the magnetic self fields can be neglected, the only equilibrium equation to solve is Poisson's equation, Eq. (3.1.13), where H, P_θ, P_z are approximated by Eqs. (3.1.19)-(3.1.21).

---

†In Eqs. (3.1.17) and (3.1.18), $V_{\alpha\theta}^0(r) \equiv \int d^3p\, v_\theta f_\alpha^0 / \int d^3p\, f_\alpha^0$ and $V_{\alpha z}^0(r) \equiv \int d^3p\, v_z f_\alpha^0 / \int d^3p\, f_\alpha^0$ are the azimuthal and axial velocity profiles associated with the *mean* motion of component α [see Eq. (1.3.20)].

3.1.3 Axisymmetric Mirror-Confined Equilibria

In this section, the equilibrium equations are obtained for the axisymmetric equilibrium configuration illustrated in Fig. 3.1.2. As in Section 3.1.2, the equilibrium is azimuthally symmetric ($\partial/\partial\theta = 0$) about the z-axis. However, since the plasma is confined axially as well as radially, equilibrium properties depend on both z and r (see Fig. 3.1.2). Therefore the equilibrium electric field, $\mathbf{E}^0(\mathbf{x}) = -\nabla\phi^0(\mathbf{x})$, produced by deviations from charge neutrality in equilibrium can be expressed as

$$\mathbf{E}^0(\mathbf{x}) = E_r^0(r, z)\hat{\mathbf{e}}_r + E_z^0(r, z)\hat{\mathbf{e}}_z, \tag{3.1.22}$$

where

$$E_r^0(r, z) = -\frac{\partial}{\partial r}\phi^0(r, z), \tag{3.1.23}$$

$$E_z^0(r, z) = -\frac{\partial}{\partial z}\phi^0(r, z), \tag{3.1.24}$$

and $\hat{\mathbf{e}}_r$ and $\hat{\mathbf{e}}_z$ are unit vectors in the r- and z-directions, respectively. In general, for the equilibrium configuration illustrated in Fig. 3.1.2, there can be an equilibrium azimuthal current, $J_\theta^0(r, z)$, associated with the mean azimuthal motions of the plasma components (e.g., a relativistic electron ring). The self magnetic field, $\mathbf{B}_0^s(\mathbf{x})$, generated by the equilibrium current $J_\theta^0(r, z)$ has both radial (r) and axial (z) components, $B_r^s(r, z)$ and $B_z^s(r, z)$. Therefore the total equilibrium magnetic field, $\mathbf{B}^0(\mathbf{x})$, can be expressed as

$$\mathbf{B}^0(\mathbf{x}) = \mathbf{B}_0^{\text{ext}}(\mathbf{x}) + B_r^s(r, z)\hat{\mathbf{e}}_r + B_z^s(r, z)\hat{\mathbf{e}}_z, \tag{3.1.25}$$

where $\mathbf{B}_0^{\text{ext}}(\mathbf{x}) = B_r^{\text{ext}}(r, z)\hat{\mathbf{e}}_r + B_z^{\text{ext}}(r, z)\hat{\mathbf{e}}_z$ is the external mirror field. In terms of the vector potentials for the external and self magnetic fields, $\mathbf{B}_0^{\text{ext}}(\mathbf{x})$ and $\mathbf{B}_0^s(\mathbf{x})$ can be represented as

$$\mathbf{B}_0^{\text{ext}}(\mathbf{x}) = \nabla \times \mathbf{A}_0^{\text{ext}}(\mathbf{x}); \quad \mathbf{B}_0^s(\mathbf{x}) = \nabla \times \mathbf{A}_0^s(\mathbf{x}), \tag{3.1.26}$$

where $\nabla = \hat{\mathbf{e}}_r(\partial/\partial r) + \hat{\mathbf{e}}_\theta(1/r)(\partial/\partial\theta) + \hat{\mathbf{e}}_z(\partial/\partial z)$, and

$$\mathbf{A}_0^{\text{ext}}(\mathbf{x}) = A_\theta^{\text{ext}}(r, z)\hat{\mathbf{e}}_\theta; \quad \mathbf{A}_0^s(\mathbf{x}) = A_\theta^s(r, z)\hat{\mathbf{e}}_\theta. \tag{3.1.27}$$

From Eqs. (3.1.26) and (3.1.27), the equilibrium self magnetic field can be expressed in terms of $A_\theta^s(r, z)$ as

$$B_z^s(r, z) = \frac{1}{r}\frac{\partial}{\partial r} r A_\theta^s(r, z), \tag{3.1.28}$$

$$B_r^s(r, z) = -\frac{\partial}{\partial z} A_\theta^s(r, z). \tag{3.1.29}$$

Since the equilibrium depends on z, there are only *two* single-particle constants of the motion for the equilibrium field configuration given in Eqs. (3.1.22) and (3.1.25). These are the total energy H and the canonical angular momentum P_θ, where

$$H \equiv (m_\alpha^2 c^4 + c^2 \mathbf{p}^2)^{1/2} + e_\alpha \phi^0(r, z) \tag{3.1.30}$$

$$P_\theta \equiv r\left[p_\theta + \frac{e_\alpha}{c} A_\theta^{\text{ext}}(r, z) + \frac{e_\alpha}{c} A_\theta^s(r, z)\right]. \tag{3.1.31}$$

Equations (3.1.30) and (3.1.31) are relativistically correct expressions for the single-particle constants. In the spirit of an equilibrium theory based on the Vlasov-Maxwell equations, any distribution function $f_\alpha^0(\mathbf{x}, \mathbf{p})$ that depends only on H and P_θ is a solution to the steady-state ($\partial/\partial t = 0$) Vlasov equation, that is, equilibrium distribution functions of the form[25]

$$f_\alpha^0(\mathbf{x}, \mathbf{p}) = f_\alpha^0(H, P_\theta) \tag{3.1.32}$$

are solutions to Eq. (1.3.8). This result can be verified by substituting Eq. (3.1.32) into Eq. (1.3.8) and making use of Eqs. (3.1.22)-(3.1.31). It is straightforward to show, by means of Eqs. (1.3.10), (1.3.11), and (3.1.22)-(3.1.29), that the equilibrium self-field potentials $\phi^0(r, z)$ and $A_\theta^s(r, z)$ are determined self-consistently in terms of $f_\alpha^0(H, P_\theta)$ from

$$\frac{1}{r}\frac{\partial}{\partial r} r \frac{\partial}{\partial r}\phi^0(r, z) + \frac{\partial^2}{\partial z^2}\phi^0(r, z) = -4\pi \sum_\alpha e_\alpha \int d^3p\, f_\alpha^0(H, P_\theta), \tag{3.1.33}$$

$$\frac{\partial}{\partial r}\frac{1}{r}\frac{\partial}{\partial r} r A_\theta^s(r, z) + \frac{\partial^2}{\partial z^2} A_\theta^s(r, z) = -\frac{4\pi}{c}\sum_\alpha e_\alpha \int d^3p\, v_\theta f_\alpha^0(H, P_\theta), \tag{3.1.34}$$

where $v_\theta \equiv (p_\theta/m_\alpha)(1 + \mathbf{p}^2/m_\alpha^2 c^2)^{-1/2}$ [see Eq. (3.1.11)]. As in Section 3.1.2, Eqs. (3.1.33) and (3.1.34) are generally *nonlinear* differential equations for the self-field potentials. This follows since H and P_θ depend on $\phi^0(r, z)$ and $A_\theta^s(r, z)$ [see Eqs. (3.1.30) and (3.1.31)].

For an axisymmetric mirror-confined equilibrium (Fig. 3.1.2), the procedure for constructing self-consistent equilibria from the steady-state Vlasov-Maxwell equations can be summarized as follows. First, specify a functional form for $f_\alpha^0(H, P_\theta)$. Second, using this form for $f_\alpha^0(H, P_\theta)$, calculate the equilibrium charge density $\rho^0(r, z)$ and the equilibrium current density $J_\theta^0(r, z)$, where†

†In Eq. (3.1.36), $V_{\alpha\theta}^0(r, z) \equiv \int d^3p\, v_\theta f_\alpha^0 / \int d^3p\, f_\alpha^0$ is the azimuthal velocity profile associated with the *mean* motion of component α [see Eq. (1.3.20)].

$$\rho^0(r,z)=\sum_\alpha e_\alpha n_\alpha^0(r,z)=\sum_\alpha e_\alpha\int d^3p\, f_\alpha^0(H,P_\theta), \tag{3.1.35}$$

$$J_\theta^0(r,z)=\sum_\alpha e_\alpha n_\alpha^0(r,z)\,V_{\alpha\theta}^0(r,z)=\sum_\alpha e_\alpha\int d^3p\, v_\theta f_\alpha^0(H,P_\theta). \tag{3.1.36}$$

Finally, making use of the expressions obtained for $\rho^0(r,z)$ and $J_\theta^0(r,z)$, calculate the equilibrium self-field potentials from Eqs. (3.1.33) and (3.1.34). As in Section 1.3.2, once $\phi^0(r,z)$ and $A_\theta^s(r,z)$ are known, other equilibrium properties can be calculated explicitly from the equilibrium distribution function $f_\alpha^0(H,P_\theta)$.

If the particle motions are nonrelativistic and the self magnetic field $\mathbf{B}_0^s(\mathbf{x})$ is negligibly small, the single-particle constants of the motion defined in Eqs. (3.1.30) and (3.1.31) can be approximated by[75]

$$H=m_\alpha c^2+\frac{\mathbf{p}^2}{2m_\alpha}+e_\alpha\phi^0(r,z), \tag{3.1.37}$$

$$P_\theta=r\left[p_\theta+\frac{e_\alpha}{c}A_\theta^{\text{ext}}(r,z)\right]. \tag{3.1.38}$$

Note that $A_\theta^s(r,z)$ has been neglected in comparison with $A_\theta^{\text{ext}}(r,z)$ in Eq. (3.1.38). Insofar as the equilibrium is electrically nonneutral and the self magnetic field can be neglected, the only equilibrium equation to solve is Poisson's equation, Eq. (3.1.33), where H and P_θ are approximated by Eqs. (3.1.37) and (3.1.38).

3.2 NONRELATIVISTIC NONDIAMAGNETIC EQUILIBRIA

3.2.1 General Discussion

In this section self-consistent Vlasov equilibria are constructed for a nonneutral plasma column aligned parallel to a uniform external magnetic field, $\mathbf{B}_0^{\text{ext}}(\mathbf{x})=B_0\hat{\mathbf{e}}_z$.[76–78] Use is made of the steady-state ($\partial/\partial t=0$) Vlasov-Maxwell description of axisymmetric column equilibria discussed in Section 3.1.2. It is assumed that external boundaries are sufficiently far removed from the plasma that their influence on the equilibrium configuration can be ignored. The equilibrium configuration and cylindrical polar coordinate system (r,θ,z) used in the present analysis are illustrated in Fig. 3.1.1. Equilibrium properties are assumed to be independent of z ($\partial/\partial z=0$) and azimuthally symmetric about an axis of symmetry parallel to $B_0\hat{\mathbf{e}}_z$. It is further assumed that the particle motions are nonrelativistic and that the axial and azimuthal self magnetic fields, $B_z^s(r)$ and $B_\theta^s(r)$, are negligibly small. Therefore the relevant equations that describe the equilibrium configuration are Eqs. (3.1.12), (3.1.13), and (3.1.19)-(3.1.21).

To simplify the analysis, it is assumed that the nonneutral plasma column is composed only of electrons, that is, no positive ions are present in the system:

$$f_i^0(\mathbf{x}, \mathbf{p}) = 0. \tag{3.2.1}$$

The analysis can be extended in a straightforward manner to include a partially neutralizing ion background. Since the particle motions are nonrelativistic, and the diamagnetic fields are assumed to be negligibly small, the equilibrium distribution function for the electrons is of the form [see Eqs. (3.1.12) and (3.1.19)–(3.1.21)]

$$f_e^0(\mathbf{x}, \mathbf{p}) = f_e^0(H, P_\theta, p_z), \tag{3.2.2}$$

where the electron energy H and canonical angular momentum P_θ are defined by†

$$H = \frac{\mathbf{p}^2}{2m_e} - e\phi^0(r), \tag{3.2.3}$$

$$P_\theta = r\left(p_\theta - m_e r \frac{\Omega_e}{2}\right). \tag{3.2.4}$$

In Eqs. (3.2.3) and (3.2.4), m_e is the electron mass, $-e$ is the electron charge, $\Omega_e = eB_0/m_e c$ is the electron cyclotron frequency, r is the radial distance from the axis of symmetry, $\phi^0(r)$ is the electrostatic potential, $\mathbf{p}$ is the mechanical momentum, and $\mathbf{p}^2 \equiv p_r^2 + p_\theta^2 + p_z^2$. For nonrelativistic, nondiamagnetic, axisymmetric column equilibria, any distribution function of the form given in Eq. (3.2.2) is a solution to the steady-state Vlasov equation for the electrons. Once the functional form of $f_e^0(H, P_\theta, p_z)$ is specified, the electrostatic potential $\phi^0(r)$ can be determined self-consistently from the equilibrium Poisson equation, Eq. (3.1.13). Since no ions are present, Eq. (3.1.13) reduces to

$$\frac{1}{r}\frac{\partial}{\partial r} r \frac{\partial}{\partial r}\phi^0(r) = 4\pi e \int d^3p\, f_e^0(H, P_\theta, p_z), \tag{3.2.5}$$

where $\int d^3p\, f_e^0(H, P_\theta, p_z)$ is related to the electron density $n_e^0(r)$ by

$$n_e^0(r) = \int d^3p\, f_e^0(H, P_\theta, p_z). \tag{3.2.6}$$

Since H depends on $\phi^0(r)$ [Eq. (3.2.3)], it follows from Eq. (3.2.6) that $n_e^0(r)$ cannot be evaluated in closed form until Eq. (3.2.5) has been solved for

---

†Without loss of generality, in defining the energy variable H in Eq. (3.2.3), the electron rest mass energy ($m_e c^2$ = const.) has been subtracted from the definition of H in Eq. (3.1.19).

$\phi^0(r)$. The detailed form of $n_e^0(r)$ of course depends on the specific choice of $f_e^0(H, P_\theta, p_z)$. Examples of self-consistent Vlasov equilibria in which the electron density assumes its maximum value *on* the axis of rotation and *off* the axis of rotation are discussed in Sections 3.2.2 and 3.2.3, respectively.

The steady-state Vlasov-Maxwell description can be used to calculate a variety of equilibrium properties. Once $\phi^0(r)$ has been determined from Eq. (3.2.5), closed expressions for macroscopic properties of the equilibrium can be calculated directly from $f_e^0(H, P_\theta, p_z)$ by taking the appropriate momentum moments. In addition to the electron density $n_e^0(r)$ [Eq. (3.2.6)], these properties include the mean equilibrium velocity $\mathbf{V}_e^0(\mathbf{x})$ of an electron fluid element [see Eq. (1.3.20)],

$$\mathbf{V}_e^0(\mathbf{x}) = \frac{m_e^{-1}\int d^3p\,\mathbf{p}\,f_e^0(H, P_\theta, p_z)}{\int d^3p\,f_e^0(H, P_\theta, p_z)}, \qquad (3.2.7)$$

and the equilibrium stress tensor $\boldsymbol{P}_e^0(\mathbf{x})$ for the electrons [see Eq. (1.3.22)],†

$$\boldsymbol{P}_e^0(\mathbf{x}) = \frac{1}{m_e}\int d^3p\left[\mathbf{p} - m_e\mathbf{V}_e^0(\mathbf{x})\right]\left[\mathbf{p} - m_e\mathbf{V}_e^0(\mathbf{x})\right] f_e^0(H, P_\theta, p_z). \qquad (3.2.8)$$

Note from Eq. (3.2.7) that any choice of equilibrium distribution function $f_e^0(H, P_\theta, p_z)$ assures that the mean radial velocity is equal to zero, that is,

$$V_{er}^0(r) = 0. \qquad (3.2.9)$$

Equation (3.2.9) follows from Eq. (3.2.7) since H is an even function of the radial momentum variable p_r. Similarly, it can be shown from Eq. (3.2.8) that the off-diagonal elements of the stress tensor that are proportional to $\int d^3p \ldots p_r f_e^0$ are identically zero, that is,

$$0 = \left[\boldsymbol{P}_e^0(\mathbf{x})\right]_{r\theta} = \left[\boldsymbol{P}_e^0(\mathbf{x})\right]_{\theta r} = \left[\boldsymbol{P}_e^0(\mathbf{x})\right]_{rz} = \left[\boldsymbol{P}_e^0(\mathbf{x})\right]_{zr}. \qquad (3.2.10)$$

3.2.2 Rigid-Rotor Equilibria

In this section, the subclass of equilibrium distribution functions that depend on H and P_θ exclusively through the linear combination, $H - \omega_e P_\theta$, where $\omega_e =$ const., is considered. In other words, the equilibrium distribution function for the electrons is assumed to be of the form[76–78]

†In Eqs. (3.2.7) and (3.2.8), the electron momentum $\mathbf{p}$ and velocity $\mathbf{v}$ are related by $\mathbf{p} = m_e\mathbf{v}$ since the particle motions are assumed to be nonrelativistic.

$$f_e^0(H, P_\theta, p_z) = f_e^0(H - \omega_e P_\theta, p_z), \tag{3.2.11}$$

where $\omega_e = \text{const}$. Making use of Eqs. (3.2.3) and (3.2.4), we can express the combination $H - \omega_e P_\theta$ as

$$\begin{aligned} H - \omega_e P_\theta &= \frac{1}{2m_e}\left[p_r^2 + p_z^2 + (p_\theta - m_e r\omega_e)^2\right] \\ &+ \frac{m_e}{2}\left[r^2(\omega_e \Omega_e - \omega_e^2) - \frac{2e}{m_e}\phi^0(r)\right]. \end{aligned} \tag{3.2.12}$$

It is straightforward to show, substituting Eq. (3.2.11) into Eq. (3.2.7) and making use of Eq. (3.2.12), that

$$V_{e\theta}^0(r) = \frac{m_e^{-1}\int d^3p\, p_\theta f_e^0(H - \omega_e P_\theta, p_z)}{\int d^3p\, f_e^0(H - \omega_e P_\theta, p_z)} = \omega_e r, \tag{3.2.13}$$

where r is the radial distance from the axis of symmetry. Therefore, for all equilibrium distribution functions of the form $f_e^0(H - \omega_e P_\theta, p_z)$, the *mean* azimuthal motion of the electrons corresponds to a rigid rotation about the axis of symmetry with angular velocity $\omega_e = \text{const}$. These equilibria are appropriately called *rigid-rotor* equilibria. Of course the equilibrium distribution function $f_e^0(H - \omega_e P_\theta, p_z)$ in general incorporates a *spread* in electron velocities relative to the mean.

Several general properties of rigid-rotor equilibria can be ascertained without specifying the functional form of $f_e^0(H - \omega_e P_\theta, p_z)$. For example, substituting Eq. (3.2.11) into Eq. (3.2.8) and making use of Eq. (3.2.12), we can show from symmetry arguments† that

$$[\mathbf{P}_e^0(\mathbf{x})]_{rr} = [\mathbf{P}_e^0(\mathbf{x})]_{\theta\theta} = \frac{1}{2m_e}\int d^3p\,[p_r^2 + (p_\theta - m_e r\omega_e)^2] f_e^0(H - \omega_e P_\theta, p_z). \tag{3.2.14}$$

Since $[\mathbf{P}_e^0(\mathbf{x})]_{r\theta} = [\mathbf{P}_e^0(\mathbf{x})]_{\theta r} = 0$ [see Eq. (3.2.10)], it follows from Eq. (3.2.14) that the particle stress tensor is isotropic in the plane perpendicular to $B_0\hat{\mathbf{e}}_z$. Therefore it is meaningful to define an effective electron temperature perpendicular to $B_0\hat{\mathbf{e}}_z$ by the relation

$$T_{e\perp}^0(r) = \frac{[\mathbf{P}_e^0(\mathbf{x})]_{rr}}{n_e^0(r)} = \frac{[\mathbf{P}_e^0(\mathbf{x})]_{\theta\theta}}{n_e^0(r)}. \tag{3.2.15}$$

†In obtaining Eq. (3.2.14) use is made of the fact that $H - \omega_e P_\theta$ is symmetric under the interchange of p_r and $p_\theta - m_e r\omega_e$ [see Eq. (3.2.12)].

Combining Eqs. (3.2.6), (3.2.14), and (3.2.15) gives

$$T_{e\perp}^{0}(r)=\frac{(2m_e)^{-1}\int d^3p\,[p_r^2+(p_\theta-m_e r\omega_e)^2]\,f_e^0(H-\omega_e P_\theta,p_z)}{\int d^3p\,f_e^0(H-\omega_e P_\theta,p_z)}. \tag{3.2.16}$$

As a further property of equilibrium distribution functions of the form $f_e^0(H-\omega_e P_\theta, p_z)$, it can be shown from Eqs. (3.2.5) and (3.2.6) that a necessary and sufficient condition for radial confinement $[n_e^0(r\to\infty)=0]$ of the electron gas is[78]

$$\omega_e\Omega_e-\omega_e^2-\frac{\omega_{pe}^2(0)}{2}>0, \tag{3.2.17}$$

where $\omega_{pe}^2(0)=4\pi n_e^0(r=0)e^2/m_e$. Equation (3.2.17) is simply a statement that magnetic restoring forces must be sufficiently strong to overcome centrifugal and electrostatic repulsive forces in order for the equilibrium to be radially confined. Whenever the inequality in Eq. (3.2.17) is satisfied, it can be shown that $n_e^0(r)$ is a nonincreasing function of r, that is, $n_e^0(r_2)\leqslant n_e^0(r_1)$ for $r_2>r_1$. For equilibrium distribution functions of the form $f_e^0(H-\omega_e P_\theta, p_z)$, the electron density profile is bell-shaped with $n_e^0(r)$ assuming its maximum value on the axis of rotation ($r=0$). The inequality in Eq. (3.2.17) can be expressed in the equivalent form

$$\omega_e^-<\omega_e<\omega_e^+, \tag{3.2.18}$$

where

$$\omega_e^{\pm}\equiv\frac{\Omega_e}{2}\left\{1\pm\left[1-\frac{2\omega_{pe}^2(0)}{\Omega_e^2}\right]^{1/2}\right\}. \tag{3.2.19}$$

If ω_e is in the range given by Eq. (3.2.18), the equilibrium density profile is radially confined with $n_e^0(r)$ decreasing to zero as $r\to\infty$. The region of parameter space $[\omega_e, 2\omega_{pe}^2(0)/\Omega_e^2]$ corresponding to radial confinement is indicated by the shaded region in Fig. 3.2.1.[78] Note that in the limit of Brillouin flow $2\omega_{pe}^2(0)/\Omega_e^2=1$, the angular velocity of rotation is $\omega_e=\omega_e^{\pm}=\Omega_e/2$.

Another interesting property of rigid-rotor equilibria can also be demonstrated. Whenever ω_e is closely tuned to the laminar rotation velocity ω_e^+ or ω_e^-, that is,

$$\omega_e=\omega_e^+(1-\delta)\quad\text{or}\quad\omega_e=\omega_e^-(1+\delta),\quad 0<\delta\ll 1, \tag{3.2.20}$$

it can be shown that the equilibrium density profile has a characteristic radial dimension R_p much larger than a thermal electron Debye length, and that the

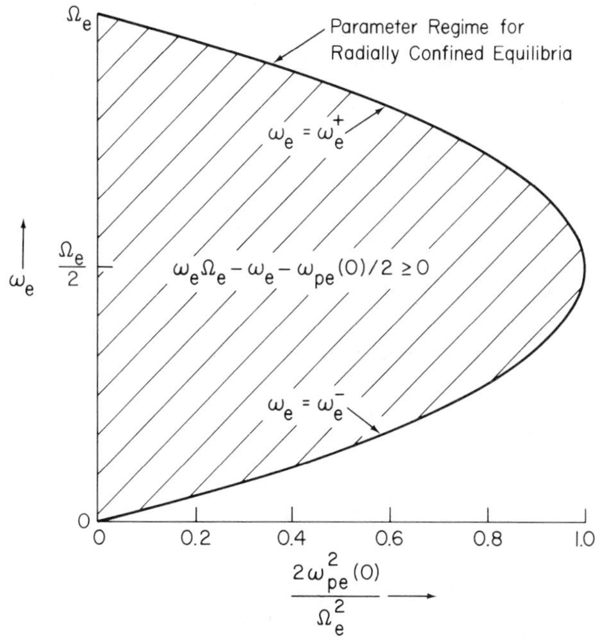

Fig. 3.2.1 For values of ω_e and $2\omega_{pe}^2(0)/\Omega_e^2$ in the shaded region of parameter space, nonneutral rigid-rotor equilibria of the form $f_e^0(H-\omega_e P_\theta, p_z)$ are radially confined, with $n_e^0(r\to\infty)=0$ [Eq. (3.2.17)].

electron density in the column interior is approximately constant with $n_e^0(r<R_p) \simeq n_e^0(0)$. For this subclass of constant-density rigid-rotor equilibria with $\omega_e \simeq \omega_e^+$ or $\omega_e \simeq \omega_e^-$, it is relatively straightforward to carry out a stability analysis based on the linearized Vlasov-Maxwell equations to determine the dispersive properties of small-amplitude body waves propagating in the column interior (see Section 3.7).

As an example of a rigid-rotor Vlasov equilibrium, consider the electron distribution function specified by[76, 77]

$$f_e^0(H-\omega_e P_\theta, p_z) = \frac{\bar{n}_e}{2\pi m_e}\,\delta(p_z - m_e V_{ez}^0) \times \delta\left(H - \omega_e P_\theta - \frac{m_e V_{ez}^{0\,2}}{2} - \frac{m_e V_{e\perp}^{0\,2}}{2}\right), \tag{3.2.21}$$

where $\bar{n}_e$, ω_e, V_{ez}^0 and $V_{e\perp}^0$ are constants, $\bar{n}_e$ is positive, and ω_e is in the interval $\omega_e^- < \omega_e < \omega_e^+$ [see Eq. (3.2.18)]. Equation (3.2.21) represents a loss-cone distribution for the electrons. Substituting Eq. (3.2.21) into Poisson's equation, and assuming that there are no external boundaries, it can be shown from Eqs. (3.2.5) and (3.2.6) that†

$$\phi^0(r) = \begin{cases} (\pi \bar{n}_e e) r^2, & 0 < r < R_p, \\ (\pi \bar{n}_e e) R_p^2 \left(1 + 2\,\ln \dfrac{r}{R_p}\right), & r > R_p, \end{cases} \tag{3.2.22}$$

and

$$n_e^0(r) = \begin{cases} \bar{n}_e, & 0 < r < R_p, \\ 0, & r > R_p, \end{cases} \tag{3.2.23}$$

where the column radius R_p is defined by

$$R_p = \frac{[V_{e\perp}^0/\omega_{pe}(0)]}{[(\omega_e - \omega_e^-)(\omega_e^+ - \omega_e)/\omega_{pe}^2(0)]^{1/2}}\,. \tag{3.2.24}$$

In Eq. (3.2.24), ω_e^- and ω_e^+ are defined in Eq. (3.2.19) and $\omega_{pe}^2(0) = 4\pi n_e^0(r=0) \times e^2/m_e = 4\pi \bar{n}_e e^2/m_e$. Note from Eq. (3.2.24) that R_p is uniquely determined in terms of properties of the equilibrium distribution function, Eq. (3.2.21). No approximations have been made in obtaining Eqs. (3.2.22)-(3.2.24) from Eqs. (3.2.5), (3.2.6), and (3.2.21). The electron density is constant and equal to $\bar{n}_e$ in the column interior, and the radial boundary at $r = R_p$ that separates the constant-density and zero-density regions is *sharp* (see Fig. 3.2.2). Note from Eq. (3.2.24) that the column radius R_p is large in comparison with the characteristic thermal Debye length, $\lambda_D = V_{e\perp}^0/\omega_{pe}(0)$, whenever $\omega_e = \omega_e^+(1-\delta)$ or $\omega_e = \omega_e^-(1+\delta)$ for $0 < \delta \ll 1$.

Other properties of the equilibrium can also be calculated from the equilibrium distribution function, Eq. (3.2.21). For example, when Eq. (3.2.21) is substituted into Eq. (3.2.7), it is readily verified that the mean axial velocity is

$$V_{ez}^0(r) = V_{ez}^0 = \text{const.} \tag{3.2.25}$$

Furthermore, substituting Eq. (3.2.21) into Eq. (3.2.8), we can show that the off-diagonal *r-z* particle stresses are identically zero [see also Eq. (3.2.10)],

†It is assumed that $\phi^0(r=0) = 0$ without loss of generality.

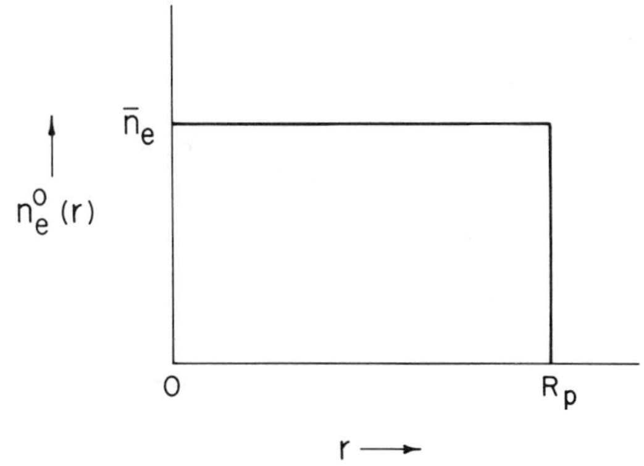

Fig. 3.2.2 Plot of $n_e^0(r)$ versus r for the equilibrium distribution function in Eq. (3.2.21). The electron density is constant inside the column [Eq. (3.2.23)], and the column radius R_p is defined in Eq. (3.2.24).

$$0 = \left[P_e^0(\mathrm{x})\right]_{rz} = \left[P_e^0(\mathrm{x})\right]_{zr}, \tag{3.2.26}$$

and that the electron motion parallel to the confining field is *cold*, that is,

$$0 = \left[P_e^0(\mathrm{x})\right]_{zz}. \tag{3.2.27}$$

The transverse temperature $T_{e\perp}^0(r)$ defined in Eq. (3.2.16), however, is nonzero for the equilibrium distribution function in Eq. (3.2.21). Substituting Eq. (3.2.21) into Eq. (3.2.16) gives

$$T_{e\perp}^0(r) = \frac{m_e V_{e\perp}^{0^2}}{2}\left(1 - \frac{r^2}{R_p^2}\right), \quad 0 < r < R_p, \tag{3.2.28}$$

where R_p is defined in Eq. (3.2.24). As illustrated in Fig. 3.2.3, the temperature profile is parabolic with $T_{e\perp}^0(r)$ assuming its maximum value $(m_e V_{e\perp}^{0^2}/2)$ at $r = 0$, and its minimum value (0) at $r = R_p$.

For the equilibrium distribution function in Eq. (3.2.21), it is straightforward to test the validity of neglecting the axial diamagnetic field $B_z^s(r)$ when the particle motions are nonrelativistic and the equilibrium azimuthal electron current is $J_\theta^0(r) = -n_e^0(r)er\omega_e$. The θ-component of the $\nabla \times \mathbf{B}_0^s$ Maxwell equation can be expressed as

$$\frac{\partial}{\partial r} B_z^s(r) = \frac{4\pi e\omega_e r n_e^0(r)}{c}. \tag{3.2.29}$$

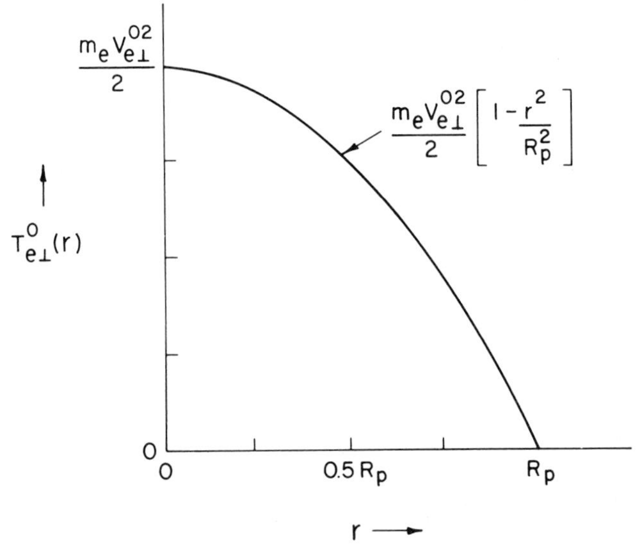

Fig. 3.2.3 Plot of $T_{e\perp}^{0}(r)$ versus r for the equilibrium distribution function in Eq. (3.2.21). The temperature profile is parabolic [Eq. (3.2.28)], assuming its maximum value at $r = 0$ and minimum value at $r = R_p$.

Since the electron density is constant in the column interior [Eq. (3.2.23)], Eq. (3.2.29) is readily integrated. The magnitude of the axial diamagnetic field assumes its maximum value at $r = 0$:

$$\left|\frac{B_z^s(r=0)}{B_0}\right| = \frac{4\pi\bar{n}_e e^2/m_e}{2\Omega_e \omega_e}\left(\frac{\omega_e R_p}{c}\right)^2 . \tag{3.2.30}$$

For a radially confined equilibrium, it follows from Eqs. (3.2.18) and (3.2.19) that $\omega_e > \omega_e^- \geqslant \omega_{pe}^2(0)/2\Omega_e$.† Therefore the coefficient multiplying $(\omega_e R_p/c)^2$ in Eq. (3.2.30) never exceeds unity. Since the particle motions are nonrelativistic, $(\omega_e R_p/c)^2 \ll 1$, and it follows from Eq. (3.2.30) that the axial diamagnetic field is negligibly small in comparison with B_0,

$$\left|\frac{B_z^s(r=0)}{B_0}\right| \ll 1. \tag{3.2.31}$$

†For low densities, $2\omega_{pe}^2(0)/\Omega_e^2 \ll 1$, note from Eq. (3.2.19) that $\omega_e^- \simeq \omega_{pe}^2(0)/2\Omega_e$.

As a further example of a rigid-rotor equilibrium, consider the electron distribution function specified by[76,77]

$$f_e^0(H-\omega_e P_\theta, p_z) = \frac{\overline{n}_e}{(2\pi m_e \Theta_e)^{3/2}} \exp\left[-\frac{H-\omega_e P_\theta}{\Theta_e}\right], \tag{3.2.32}$$

where $\overline{n}_e$ and Θ_e are positive constants, and ω_e is in the interval $\omega_e^- < \omega_e < \omega_e^+$ [see Eq. (3.2.18)]. The Gibbs distribution function in Eq. (3.2.32) represents the thermal equilibrium distribution to which an isolated electron gas column would evolve through binary collisions. (Keep in mind, however, that binary collisions are not included in the Vlasov-Maxwell description.) Since H is an even function of the axial momentum variable p_z, it follows from Eqs. (3.2.7) and (3.2.32) that the mean axial velocity of the electrons is equal to zero:

$$V_{ez}^0(r) = 0. \tag{3.2.33}$$

Furthermore, it follows from Eqs. (3.2.8) and (3.2.32) that the equilibrium stress tensor for the electrons can be expressed as

$$\mathbf{P}_e^0(\mathbf{x}) = n_e^0(r)\Theta_e(\hat{\mathbf{e}}_r\hat{\mathbf{e}}_r + \hat{\mathbf{e}}_\theta\hat{\mathbf{e}}_\theta + \hat{\mathbf{e}}_z\hat{\mathbf{e}}_z), \tag{3.2.34}$$

where $\hat{\mathbf{e}}_r$, $\hat{\mathbf{e}}_\theta$, and $\hat{\mathbf{e}}_z$ are unit vectors in the r-, θ- and z-directions, respectively (see Fig. 3.1.1), and $n_e^0(r) = \int d^3p\, f_e^0(H-\omega_e P_\theta, p_z)$ is the electron density. It is evident from Eq. (3.2.34) that the equilibrium stress tensor is isotropic, and that Θ_e can be identified with the electron temperature. Substituting Eq. (3.2.32) into Eq. (3.2.6) gives

$$n_e^0(r) = \overline{n}_e \exp\left\{-\frac{m_e}{2\Theta_e}\left[r^2(\omega_e\Omega_e - \omega_e^2) - \frac{2e}{m_e}\phi^0(r)\right]\right\}, \tag{3.2.35}$$

and Poisson's equation, Eq. (3.2.5), can be expressed as

$$\frac{1}{r}\frac{\partial}{\partial r} r \frac{\partial}{\partial r}\phi^0(r) = 4\pi e \overline{n}_e \exp\left\{-\frac{m_e}{2\Theta_e}\left[r^2(\omega_e\Omega_e - \omega_e^2) - \frac{2e}{m_e}\phi^0(r)\right]\right\}. \tag{3.2.36}$$

Without loss of generality it is assumed that $\phi^0(r=0)=0$. Therefore $\overline{n}_e$ can be identified with the electron density on axis, that is, $n_e^0(r=0) = \overline{n}_e$ [see Eq. (3.2.35)]. Equations (3.2.35) and (3.2.36) cannot be solved in closed form for $n_e^0(r)$ and $\phi^0(r)$. However, Eq. (3.2.36) can be integrated numerically. For

$\omega_e/\Omega_e = 0.165$ and $2\omega_{pe}^2(0)/\Omega_e^2 = 0.5$, the electron density profile has the form shown in Fig. 3.2.4. Note that $n_e^0(r)$ is bell-shaped with the electron density assuming its maximum value at $r = 0$. For the parameters chosen in Fig. 3.2.4, the characteristic radius R_p of the electron gas column is of the order of four thermal Debye lengths, that is,

$$R_p \approx 4\lambda_D = 4\left(\frac{\Theta_e}{4\pi\bar{n}_e e^2}\right)^{1/2} . \tag{3.2.37}$$

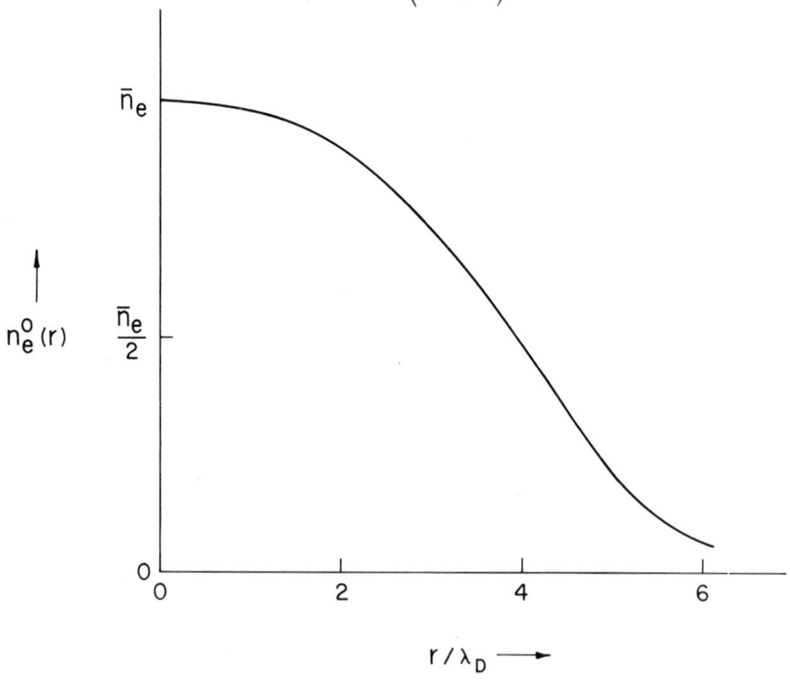

Fig. 3.2.4 Plot of $n_e^0(r)$ versus r/λ_D for the equilibrium distribution function in Eq. (3.2.32). This profile is obtained by numerically integrating Eq. (3.2.36) for $\omega_e/\Omega_e = 0.165$ and $2\omega_{pe}^2(0)/\Omega_e^2 = 0.5$, and substituting the resulting expression for $\phi^0(r)$ into Eq. (3.2.35).

It should be emphasized, however, that if ω_e is closely tuned to the laminar rotation velocity, ω_e^+ or ω_e^-, defined in Eq. (3.2.19), the equilibrium density profile determined from Eqs. (3.2.35) and (3.2.36) is radially very broad in units of λ_D. Although the details will not be presented here, it can be shown from Eqs. (3.2.35) and (3.2.36) that, when $\omega_e = \omega_e^+(1-\delta)$ or $\omega_e = \omega_e^-(1+\delta)$, for $0<\delta \ll 1$,

$$n_e^0(r) \simeq \begin{cases} \bar{n}_e, & 0<r<R_p, \\ 0, & r>R_p, \end{cases} \tag{3.2.38}$$

where

$$R_p \approx \left(\frac{\Theta_e}{4\pi\bar{n}_e e^2}\right)^{1/2} \ell n \left\{\left[\ell n\left(1+\frac{(\omega_e^+-\omega_e)(\omega_e-\omega_e^-)}{\omega_{pe}^2(0)/2}\right)\right]^{-1}\right\}. \tag{3.2.39}$$

For $\delta \ll 1$, it follows from Eq. (3.2.39) that $R_p \gg \lambda_D \equiv (\Theta_e/4\pi\bar{n}_e e^2)^{1/2}$. The electron density $n_e^0(r)$ drops to zero rapidly for $r > R_p$, with a thin surface thickness ($\approx \lambda_D$).

3.2.3 Hollow Beam Equilibria with Shear in Angular Velocity

Hollow beam equilibria in which the electron density assumes its maximum value off the axis of rotation ($r = 0$) have received considerable attention in the literature, principally in connection with the diocotron instability (see Sections 2.2 and 2.10). These equilibria are characterized by a shear in the angular velocity of mean rotation, that is, $\partial\omega_e(r)/\partial r \neq 0$, where

$$\omega_e(r) \equiv \frac{V_{e\theta}^0(r)}{r} = \frac{m_e^{-1}\int d^3p\, p_\theta f_e^0(H, P_\theta, p_z)}{\int d^3p f_e^0(H, P_\theta, p_z)}. \tag{3.2.40}$$

In this section, two examples of hollow beam equilibria with shear in angular velocity are briefly examined. Use is made of the nonrelativistic, nondiamagnetic Vlasov description discussed in Section 3.2.1.

As a first example of a hollow beam equilibrium, consider the electron distribution function specified by[77, 109]

$$f_e^0(H, P_\theta, p_z) = \frac{\bar{n}_e R_0}{2\pi m_e}\,\delta(P_\theta - P_0)\,\delta(H - H_0), \tag{3.2.41}$$

where $\bar{n}_e$, R_0, H_0, and P_0 are positive constants, and $H_0 > P_0\Omega_e$, by hypothesis. When Eq. (3.2.41) is substituted into Eq. (3.2.6), it is straightforward to show that the electron density profile is

$$n_e^0(r) = \begin{cases} 0, & 0 < r < R_0, \\ \bar{n}_e \dfrac{R_0}{r}, & R_0 < r < R_p, \\ 0, & r > R_p, \end{cases} \tag{3.2.42}$$

where R_0 and R_p are the extremes of the interval on which the inequality,

$$H_0 + e\phi^0(r) - \frac{m_e}{2}\left(\frac{r\Omega_e}{2} + \frac{P_0}{m_e r}\right)^2 \geqslant 0, \tag{3.2.43}$$

is satisfied. On the assumption that there are no external boundaries, the equilibrium electrostatic potential $\phi^0(r)$ for this example can be expressed as

$$\phi^0(r) = \begin{cases} 0, & 0 < r < R_0, \\ 4\pi e \bar{n}_e R_0 \left(r - R_0 - R_0 \ln \dfrac{r}{R_0}\right), & R_0 < r < R_p, \\ 4\pi e \bar{n}_e R_0 \left(R_p - R_0 - R_0 \ln \dfrac{R_p}{R_0} \right. & \\ \qquad \left. - (R_p - R_0) \ln \dfrac{r}{R_p}\right), & r > R_p. \end{cases} \tag{3.2.44}$$

Note from Eq. (3.2.42) that the electron density falls off as $1/r$ in the range $R_0 < r < R_p$, and that the radial boundaries of the column (at $r = R_0$ and $r = R_p$) are *sharp* (see Fig. 3.2.5). Equations (3.2.43) and (3.2.44) can be combined to determine R_0 and R_p. For example, since $\phi^0(R_0) = 0$, Eq. (3.2.43) gives (for $H_0 > \Omega_e P_0 > 0$)

$$R_0 = \left(\frac{2}{m_e \Omega_e^2}\right)^{1/2} \left[H_0^{1/2} - (H_0 - \Omega_e P_0)^{1/2}\right]. \tag{3.2.45}$$

A closed expression for R_p is not accessible analytically.† However, once the values of H_0, P_0, Ω_e, etc., are specified, R_p can be determined numerically from Eqs. (3.2.43) and (3.2.44).

---

† If the equilibrium layer of electrons is thin with $\Delta = R_p \; R_0 \ll R_0$, it is straightforward to obtain an approximate expression for $R_p = R_0 + \Delta$ from Eqs. (3.2.43) and (3.2.44).

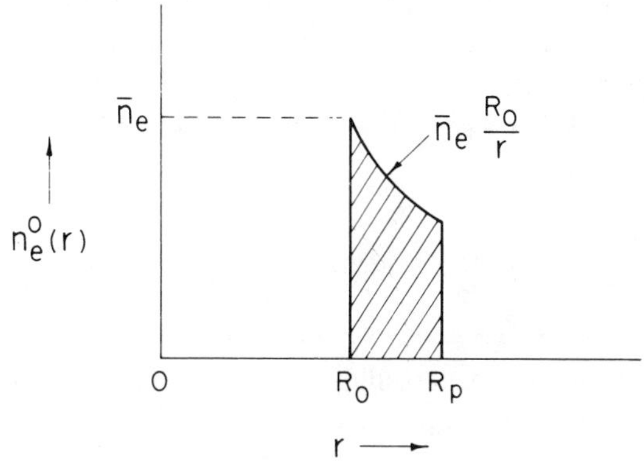

Fig. 3.2.5 Plot of $n_e^0(r)$ versus r for the equilibrium distribution function in Eq. (3.2.41). The electron density profile is hollow [Eq. (3.2.42)] with inner radius R_0 and outer radius R_p determined from the extremes of the interval on which the inequality in Eq. (3.2.43) is satisfied.

Other properties of the equilibrium can also be calculated from the equilibrium distribution function, Eq. (3.2.41). For example, when Eq. (3.2.41) is substituted into Eq. (3.2.7), it is readily verified that the mean axial velocity is equal to zero, $V_{ez}^0(r) = 0$, and that the angular velocity of mean rotation is given by

$$\omega_e(r) = \frac{V_{e\theta}^0(r)}{r} = \frac{\Omega_e}{2} + \frac{P_0}{m_e r^2}, \quad R_0 < r < R_p. \tag{3.2.46}$$

Note from Eq. (3.2.46) that the mean azimuthal motion of the equilibrium does *not* correspond to a rigid rotation about the axis of symmetry, that is, $\partial\omega_e(r)/\partial r \neq 0$. Furthermore, when Eq. (3.2.41) is substituted into Eq. (3.2.8), it can be shown that the equilibrium stress tensor is given by

$$\boldsymbol{P}_e^0(\mathbf{x}) = n_e^0(r)\left[H_0 + e\phi^0(r) - \frac{m_e}{2}\left(\frac{r\Omega_e}{2} + \frac{P_0}{m_e r}\right)^2\right](\hat{\mathbf{e}}_r\hat{\mathbf{e}}_r + \hat{\mathbf{e}}_z\hat{\mathbf{e}}_z), \tag{3.2.47}$$

where $n_e^0(r)$ and $\phi^0(r)$ are defined in Eqs. (3.2.42) and (3.2.44), and $\hat{\mathbf{e}}_r$ and $\hat{\mathbf{e}}_z$ are unit vectors in the r- and z-directions, respectively (see Fig. 3.1.1).

The quantity in square brackets in Eq. (3.2.47) is equal to zero at $r = R_0$ and $r = R_p$ [see Eq. (3.2.43)]. Therefore the particle stress tensor for the electrons vanishes identically at the inner and outer surfaces of the column. Note from Eq. (3.2.47) that the effective axial and radial temperatures for $R_0 < r < R_p$ can be expressed as

$$T_{ez}^0(r) = T_{er}^0(r) = H_0 + e\phi^0(r) - \frac{m_e}{2}\left(\frac{r\Omega_e}{2} + \frac{P_0}{m_e r}\right)^2 . \qquad (3.2.48)$$

As an example of a hollow beam equilibrium with *diffuse* radial boundaries, consider the electron distribution function specified by[77]

$$f_e^0(H, P_\theta, p_z) = \frac{AP_\theta^2}{(2\pi m_e\,\Theta_e)^{3/2}} \exp\left[-\frac{H - \bar{\omega}_e P_\theta}{\Theta_e}\right], \qquad (3.2.49)$$

where A, $\bar{\omega}_e$, and Θ_e are positive constants. When Eq. (3.2.49) is substituted into Eq. (3.2.6), it is straightforward to show that the electron density profile is

$$n_e^0(r) = Am_e^2 r^2\left[\frac{\Theta_e}{m_e} + r^2\left(\bar{\omega}_e - \frac{\Omega_e}{2}\right)^2\right]$$

$$\times \exp\left\{-\frac{m_e}{2\Theta_e}\left[(\bar{\omega}_e\Omega_e - \bar{\omega}_e^2)r^2 - \frac{2e}{m_e}\phi^0(r)\right]\right\}, \qquad (3.2.50)$$

where $\phi^0(r)$ is determined self-consistently from Poisson's equation,

$$\frac{1}{r}\frac{\partial}{\partial r}r\frac{\partial}{\partial r}\phi^0(r) = 4\pi e A m_e^2 r^2\left[\frac{\Theta_e}{m_e} + r^2\left(\bar{\omega}_e - \frac{\Omega_e}{2}\right)^2\right]$$

$$\times \exp\left\{-\frac{m_e}{2\Theta_e}\left[(\bar{\omega}_e\Omega_e - \bar{\omega}_e^2)r^2 - \frac{2e}{m_e}\phi^0(r)\right]\right\}. \qquad (3.2.51)$$

Equations (3.2.50) and (3.2.51) cannot be solved in closed form for $n_e^0(r)$ and $\phi^0(r)$. However, for an appropriate range of $\bar{\omega}_e$ it can be shown that the equilibrium is radially confined. The electron density $n_e^0(r)$ begins at zero when $r = 0$, increases monotonically to some maximum value $\bar{n}_{eM}$ at $r = r_M$, and subsequently decreases monotonically to zero as $r \to \infty$. Furthermore, when Eq. (3.2.49) is substituted into Eq. (3.2.7), it can be shown that the mean axial velocity is equal to zero, $V_{ez}^0(r) = 0$, and that the angular velocity of mean rotation is given by

$$\omega_e(r) = \frac{V^0_{e\theta}(r)}{r} = \overline{\omega}_e \left[1 + \frac{\dfrac{2\Theta_e}{m_e} \dfrac{\overline{\omega}_e - \Omega_e/2}{\overline{\omega}_e}}{\dfrac{\Theta_e}{m_e} + r^2 \left(\overline{\omega}_e - \dfrac{\Omega_e}{2} \right)^2} \right]. \tag{3.2.52}$$

Note from Eq. (3.2.52) that the mean azimuthal motion of the equilibrium does *not* correspond to a rigid rotation about the axis of symmetry, unless $\overline{\omega}_e = \Omega_e/2$.

3.3 RELATIVISTIC E-LAYER EQUILIBRIA

3.3.1 General Discussion

Solutions to the steady-state ($\partial/\partial t = 0$) Vlasov-Maxwell equations that correspond to relativistic E-layers[112–119] (cylindrical layers of electrons) are of considerable practical interest for Astron-like configurations.[50,51,120,121] In this section examples of self-consistent Vlasov equilibria are considered for an E-layer aligned parallel to a uniform external magnetic field, $\mathbf{B}_0^{\text{ext}}(\mathbf{x}) = B_0\hat{\mathbf{e}}_z$. Use is made of the steady-state Vlasov-Maxwell description of axisymmetric column equilibria discussed in Section 3.1.2. The equilibrium configuration and

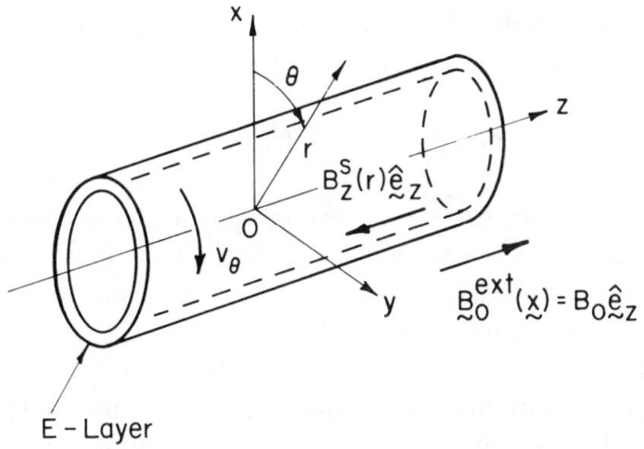

Fig. 3.3.1 Axisymmetric E-layer aligned parallel to a uniform external magnetic field, $\mathbf{B}_0^{\text{ext}}(\mathbf{x}) = B_0\hat{\mathbf{e}}_z$. Cylindrical polar coordinates (r, θ, z) are introduced with the z-axis coinciding with the axis of symmetry; θ is the polar angle in the x-y plane, and $r = \sqrt{x^2 + y^2}$ is the radial distance from the z-axis.

cylindrical polar coordinate system (r, θ, z) used in the present analysis are illustrated in Fig. 3.3.1. Equilibrium properties are assumed to be independent of z $(\partial/\partial z = 0)$ and azimuthally symmetric $(\partial/\partial\theta = 0)$ about an axis of symmetry parallel to $B_0\hat{\mathbf{e}}_z$. The mean azimuthal motion of the electrons composing the E-layer is assumed to be relativistic, and the axial self magnetic field $B_z^s(r)$ is included in the equilibrium analysis.† The relevant equations that describe the equilibrium configuration are Eqs. (3.1.8)-(3.1.15).

To simplify the theoretical analysis the following assumptions are made:

1. The E-layer is immersed in a neutralizing plasma background that provides local charge neutrality, $\rho^0(r) = 0$. Therefore, from Eqs. (3.1.13) and (3.1.16), the equilibrium radial electric field is equal to zero, $E_r^0(r) = -\partial\phi^0(r)/\partial r = 0$, and

$$\phi^0(r) = 0 \tag{3.3.1}$$

   without loss of generality. It is further assumed that the current carried by the background plasma is equal to zero. Therefore the self magnetic field is generated entirely by the E-layer.
2. External boundaries are sufficiently far removed from the E-layer that their influence on the equilibrium configuration can be ignored.
3. The equilibrium distribution function for the electrons composing the E-layer does not depend on the axial canonical momentum P_z defined in Eq. (3.1.10). In other words, the present analysis is restricted to the class of equilibrium distriubtion functions of the form

$$f_e^0(\mathbf{x}, \mathbf{p}) = f_e^0(H, P_\theta), \tag{3.3.2}$$

   where H and P_θ are the single-particle constants of the motion defined in Eqs. (3.1.8) and (3.1.9). For equilibrium distribution functions of the form $f_e^0(H, P_\theta)$, there is no axial motion of the electron beam, and the azimuthal self magnetic field is equal to zero.

As discussed in Section 3.1.2, any distribution function that is a function only of the single-particle constants of the motion in the equilibrium fields satisfies the steady-state Vlasov equation. For axisymmetric column equilibria, the total energy H and canonical angular momentum P_θ of an electron can be expressed as [see Eqs. (3.1.8), (3.1.9), and (3.3.1)]

---

†The equilibrium examples considered in Section 3.2.3 also correspond to E-layers. However, the particle motions were assumed to be nonrelativistic, and the axial self magnetic field was neglected in the analysis.

$$H = (m_e^2 c^4 + c^2 \mathbf{p}^2)^{1/2}, \tag{3.3.3}$$

$$P_\theta = r\left[p_\theta - m_e r \frac{\Omega_e}{2} - \frac{e}{c} A_\theta^s(r)\right], \tag{3.3.4}$$

where $\Omega_e = eB_0/m_e c$, m_e is the electron mass, $-e$ is the electron charge, r is the radial distance from the axis of symmetry, $A_\theta^s(r)$ is the θ-component of the equilibrium vector potential for the axial self magnetic field $B_z^s(r)$ [see Eq. (3.1.6)], $\mathbf{p}$ is the mechanical momentum, and $\mathbf{p}^2 = p_r^2 + p_\theta^2 + p_z^2$. In reducing Eq. (3.1.9) to the form given in Eq. (3.3.4), use has been made of $A_\theta^{ext}(r) = rB_0/2$.

The simplification provided by assumption 3 can be demonstrated as follows. Note from Eq. (3.3.3) that H is an even function of p_z. Therefore $f_e^0(H, P_\theta)$ is also an even function of p_z, and the mean axial velocity of the E-layer is equal to zero, that is,

$$V_{ez}^0(r) = \frac{\int d^3p\, v_z f_e^0(H, P_\theta)}{\int d^3p f_e^0(H, P_\theta)} = 0, \tag{3.3.5}$$

where $v_z = (p_z/m_e)(1 + \mathbf{p}^2/m_e^2c^2)^{-1/2}$.† Since the background plasma current is zero (assumption 1), it follows from Eqs. (3.1.7), (3.1.15), and (3.3.5) that no azimuthal self magnetic field is produced by the relativistic E-layer, that is,

$$B_\theta^s(r) = 0, \tag{3.3.6}$$

for the class of equilibrium distribution functions of the form $f_e^0(H, P_\theta)$.

Once the functional form of $f_e^0(H, P_\theta)$ is specified, the self-field vector potential $A_\theta^s(r)$ can be determined self-consistently from the θ-component of the $\nabla \times \mathbf{B}_0^s$ Maxwell equation, Eq. (3.1.14). Since the background plasma current is zero (assumption 1), Eq. (3.1.14) reduces to

$$\frac{\partial}{\partial r}\frac{1}{r}\frac{\partial}{\partial r} rA_\theta^s(r) = \frac{4\pi e}{c}\int d^3p\; v_\theta f_e^0(H, P_\theta), \tag{3.3.7}$$

where $v_\theta \equiv (p_\theta/m_e)(1 + \mathbf{p}^2/m_e^2c^2)^{-1/2}$. In Eq. (3.3.7), $-e \int d^3p\; v_\theta f_e^0(H, P_\theta)$ is related to the azimuthal electron current $J_\theta^0(r)$ by

† Similarly, $f_e^0(H, P_\theta)$ is an even function of p_r, and the mean radial velocity is equal to zero

$$V_{er}^0(r) = \int d^3p\, v_r f_e^0(H, P_\theta)/\int d^3p\, f_e^0(H, P_\theta) = 0.$$

$$J_\theta^0(r) = -en_e^0(r)V_{e\theta}^0(r) = -e\int d^3p\, v_\theta f_e^0(H, P_\theta)\,, \tag{3.3.8}$$

where $V_{e\theta}^0(r)$ is the mean azimuthal velocity of an electron fluid element,

$$V_{e\theta}^0(r) = \frac{\int d^3p v_\theta f_e^0(H, P_\theta)}{\int d^3p f_e^0(H,P_\theta)}\,, \tag{3.3.9}$$

and $n_e^0(r)$ is the electron density,

$$n_e^0(r) = \int d^3p\, f_e^0(H, P_\theta)\,. \tag{3.3.10}$$

Since P_θ depends on $A_\theta^s(r)$, Eq. (3.3.7) is generally a nonlinear differential equation for the self-field vector potential. Once the functional form of $f_e^0(H, P_\theta)$ is specified, however, $A_\theta^s(r)$ can, in principle, be determined from Eq. (3.3.7), and then the axial self magnetic field obtained from

$$B_z^s(r) = \frac{\partial}{\partial r} A_\theta^s(r) + \frac{A_\theta^s(r)}{r}\,. \tag{3.3.11}$$

3.3.2 E-Layer Equilibrium with Sharp Radial Boundaries

There is considerable latitude in the choice of equilibrium distribution function $f_e^0(H, P_\theta)$. As an example that corresponds to a relativistic E-layer with sharp radial boundaries and also permits an exact analytic determination of the axial self magnetic field $B_z^s(r)$, consider the equilibrium distribution function specified by[114]

$$f_e^0(H, P_\theta) = \frac{\overline{n}_e R_0}{2\pi\gamma_0 m_e}\,\delta(P_\theta - P_0)\,\delta(H - \gamma_0 m_e c^2)\,, \tag{3.3.12}$$

where H and P_θ are defined in Eqs. (3.3.3) and (3.3.4), P_0 = const., and $\overline{n}_e$, R_0, and γ_0 are positive constants. For the equilibrium distribution function defined in Eq. (3.3.12), all of the electrons composing the E-layer have the same value of canonical angular momentum ($P_\theta = P_0$) and the same value of energy ($H = \gamma_0 m_e c^2$). Except for the definitions of H and P_θ, Eq. (3.3.12) is similar in form to the equilibrium distribution function, Eq. (3.2.41), analyzed in Section 3.2. When Eq. (3.3.12) is substituted into Eq. (3.3.10), it is straightforward to show that the electron density profile for the E-layer is[114]

$$n_e^0(r) = \begin{cases} 0, & 0 < r < R_0\,, \\ \bar{n}_e \dfrac{R_0}{r}, & R_0 < r < R_p\,, \\ 0, & r > R_p\,, \end{cases} \tag{3.3.13}$$

where R_0 and R_p are the extremes of the interval on which the inequality,

$$\left\{ m_e^2 c^4 + c^2 \left[\frac{P_0}{r} + m_e r \frac{\Omega_e}{2} + \frac{e}{c} A_\theta^s(r) \right]^2 \right\}^{1/2} - \gamma_0 m_e c^2 \leqslant 0, \tag{3.3.14}$$

is satisfied. Note from Eq. (3.3.13) that the density profile has sharp radial boundaries (at R_0 and R_p) and is similar in form to the density profile illustrated in Fig. 3.2.5. When Eq. (3.3.12) is substituted into Eq. (3.3.8), the azimuthal current density can be expressed as[114]

$$J_\theta^0(r) = \begin{cases} 0, & 0 < r < R_0\,, \\ -\dfrac{\bar{n}_e e}{\gamma_0 m_e} \dfrac{R_0}{r} \left[\dfrac{P_0}{r} + m_e r \dfrac{\Omega_e}{2} + \dfrac{e}{c} A_\theta^s(r) \right], & R_0 < r < R_p\,, \\ 0, & r > R_p\,, \end{cases} \tag{3.3.15}$$

and Eq. (3.3.7) reduces to

$$\frac{\partial}{\partial r} \frac{1}{r} \frac{\partial}{\partial r} r A_\theta^s(r) = \begin{cases} 0, & 0 < r < R_0, \\ \dfrac{4\pi \bar{n}_e e}{\gamma_0 m_e c} \dfrac{R_0}{r} \left[\dfrac{P_0}{r} + m_e r \dfrac{\Omega_e}{2} + \dfrac{e}{c} A_\theta^s(r) \right], & R_0 < r < R_p, \\ 0, & r > R_p\,. \end{cases} \tag{3.3.16}$$

From Eqs. (3.3.8), (3.3.13), and (3.3.15), the azimuthal velocity profile for electrons composing the E-layer is

$$V_{e\theta}^0(r) = \frac{1}{\gamma_0 m_e} \left[\frac{P_0}{r} + m_e r \frac{\Omega_e}{2} + \frac{e}{c} A_\theta^s(r) \right] \tag{3.3.17}$$

for $R_0 < r < R_p$.

Note from Eq. (3.3.14) that R_0 and R_p cannot be evaluated explicitly until $A_\theta^s(r)$ has been determined from Eq. (3.3.16). Furthermore, the solution for $A_\theta^s(r)$ depends on R_0 and R_p. Therefore the condition that determines R_0 and R_p is, in effect, *nonlinear*. Keeping in mind that the self magnetic field is zero outside the E-layer ($r > R_p$), we can express the solution to Eq. (3.3.16) as

$$A_\theta^s(r) = \begin{cases} B_i^s \dfrac{r}{2}, & 0 < r < R_0, \\ aI_2\,[(r/\delta)^{1/2}] + bK_2\,[(r/\delta)^{1/2}] - \dfrac{cP_0}{er} - \dfrac{rB_0}{2}, & R_0 < r < R_p, \\ \dfrac{d}{r}, & r > R_p, \end{cases} \tag{3.3.18}$$

where B_i^s, a, b, and d are constants, I_n and K_n are modified Bessel functions of order n, and δ is defined by

$$\delta = \frac{c^2\gamma_0}{4\bar{\omega}_{pe}^2 R_0}, \tag{3.3.19}$$

where $\bar{\omega}_{pe}^2 = 4\pi\bar{n}_e e^2/m_e$. From Eqs. (3.3.11) and (3.3.18), the axial self magnetic field $B_z^s(r)$ is

$$B_z^s(r) = \begin{cases} B_i^s, & 0 < r < R_0, \\ \dfrac{1}{2(r\delta)^{1/2}}\left\{aI_1\,[(r/\delta)^{1/2}] - bK_1\,[(r/\delta)^{1/2}]\right\} - B_0, & R_0 < r < R_p, \\ 0, & r > R_p. \end{cases} \tag{3.3.20}$$

The continuity of $A_\theta^s(r)$ and $B_z^s(r)$ at $r = R_0$ and $r = R_p$ provides the four boundary conditions that determine the constants B_i^s, a, b, and d, in terms of δ, P_0, B_0, R_0, and R_p. From Eqs. (3.3.18) and (3.3.20), the continuity of $A_\theta^s(r)$ and $B_z^s(r)$ at $r = R_0$ gives

$$B_i^s \frac{R_0}{2} = aI_2\,[(R_0/\delta)^{1/2}] + bK_2\,[(R_0/\delta)^{1/2}] - \frac{cP_0}{eR_0} - \frac{R_0 B_0}{2}, \tag{3.3.21}$$

$$B_i^s = \frac{1}{2(R_0\delta)^{1/2}}\left\{aI_1[(R_0/\delta)^{1/2}] - bK_1[(R_0/\delta)^{1/2}]\right\} - B_0. \tag{3.3.22}$$

Moreover, from Eq. (3.3.20), the continuity of $B_z^s(r)$ at $r = R_p$ gives

$$\frac{1}{2(R_p\delta)^{1/2}}\left\{aI_1[(R_p/\delta)^{1/2}] - bK_1[(R_p/\delta)^{1/2}]\right\} - B_0 = 0. \tag{3.3.23}$$

The continuity of $A_\theta^s(r)$ at $r = R_p$ serves only to evaluate the constant d and is not required to determine the self magnetic field $B_z^s(r)$ [see Eq. (3.3.20)]. From Eqs. (3.3.21)-(3.3.23), the constants B_i^s, a, and b are

$$\begin{aligned} B_i^s = -B_0 - \Delta^{-1}\Bigg(&\frac{cP_0/eR_0}{4(R_0\delta)^{1/2}(R_p\delta)^{1/2}}\Big\{I_1[(R_0/\delta)^{1/2}]K_1[(R_p/\delta)^{1/2}] \\ &- I_1[(R_p/\delta)^{1/2}]K_1[(R_0/\delta)^{1/2}]\Big\} \\ &+ \frac{B_0}{2(R_0\delta)^{1/2}}\Big\{I_1[(R_0/\delta)^{1/2}]K_2[(R_0/\delta)^{1/2}] \\ &+ I_2[(R_0/\delta)^{1/2}]K_1[(R_0/\delta)^{1/2}]\Big\}\Bigg), \end{aligned} \tag{3.3.24}$$

$$\begin{aligned} a = -\Delta^{-1}\Bigg(&\frac{cP_0/eR_0}{2(R_p\delta)^{1/2}}K_1[(R_p/\delta)^{1/2}] \\ &+ B_0\Big\{\frac{R_0}{4(R_0\delta)^{1/2}}K_1[(R_0/\delta)^{1/2}] + K_2[(R_0/\delta)^{1/2}]\Big\}\Bigg), \end{aligned} \tag{3.3.25}$$

and

$$\begin{aligned} b = -\Delta^{-1}\Bigg(&\frac{cP_0/eR_0}{2(R_p\delta)^{1/2}}I_1[(R_p/\delta)^{1/2}] \\ &+ B_0\Big\{\frac{R_0}{4(R_0\delta)^{1/2}}I_1[(R_0/\delta)^{1/2}] - I_2[(R_0/\delta)^{1/2}]\Big\}\Bigg), \end{aligned} \tag{3.3.26}$$

where

$$\Delta \equiv \frac{R_0}{8(R_0\delta)^{1/2}(R_p\delta)^{1/2}}\Big\{K_1[(R_p/\delta)^{1/2}]\,I_1\,[(R_0/\delta)^{1/2}] - I_1\,[(R_p/\delta)^{1/2}]\,K_1\,[(R_0/\delta)^{1/2}]\Big\} - \frac{1}{2(R_p\delta)^{1/2}}\Big\{K_1\,[(R_p/\delta)^{1/2}]\,I_2\,[(R_0/\delta)^{1/2}] + I_1\,[(R_p/\delta)^{1/2}]\,K_2\,[(R_0/\delta)^{1/2}]\Big\}\ . \tag{3.3.27}$$

When Eqs. (3.3.24)-(3.3.27) are used, Eq. (3.3.20) constitutes a closed expression for the axial self magnetic field $B_z^s(r)$ in terms of δ, P_0, B_0, R_0, and R_p.

The values of R_0 and R_p are determined from the extremes of the interval on which the inequality in Eq. (3.3.14) is satisfied. For example, substituting Eq. (3.3.18) into Eq. (3.3.14) and making use of Eq. (3.3.21), we determine the inner radius R_0 from

$$\frac{P_0}{R_0} + \frac{eR_0}{2c}(B_0 + B_i^s) = \pm(\gamma_0^2 - 1)^{1/2}\, m_e c\ , \tag{3.3.28}$$

where B_i^s is defined in Eq. (3.3.24), and the upper (lower) sign in Eq. (3.3.28) holds if $V_{e\theta}^0(R_0)$ is positive (negative) [see Eq. (3.3.17)]. Similarly, substituting Eq. (3.3.18) into Eq. (3.3.14), we determine the outer radius R_p from

$$\frac{e}{c}\Big\{aI_2\,[(R_p/\delta)^{1/2}] + bK_2\,[(R_p/\delta)^{1/2}]\Big\} = \pm(\gamma_0^2 - 1)^{1/2} m_e c, \tag{3.3.29}$$

where a and b are defined in Eqs. (3.3.25) and (3.3.26). Since B_i^s, a and b are transcendental functions of R_0 and R_p, the inner and outer radii must be evaluated numerically from Eqs. (3.3.28) and (3.3.29). In principle, however, Eqs. (3.3.28) and (3.3.29) determine R_0 and R_p in terms of B_0 and the parameters that characterize the equilibrium distribution function defined in Eq. (3.3.12) (e.g., $\bar{n}_e$, P_0 and $\gamma_0 m_e c^2$).

As an example, consider the choice of equilibrium parameters related by

$$P_0 = \frac{5}{2}(\gamma_0^2 - 1)^{1/2}\frac{m_e c^2}{\bar{\omega}_{pe}}\ , \tag{3.3.30}$$

$$(\gamma_0^2 - 1)^{1/2} = \frac{4}{3}\Omega_e/\omega_{pe}\ , \tag{3.3.31}$$

where $\bar{\omega}_{pe}^2 = 4\pi\bar{n}_e e^2/m_e$ and $\Omega_e = eB_0/m_e c$. Using the upper sign in Eq.

(3.3.28), it can be shown from Eqs. (3.3.24)-(3.3.29) that $B_i^s = -1.33B_0$, $R_0 = 2c/\bar{\omega}_{pe}$, and $R_p \simeq 3.65c/\bar{\omega}_{pe}$.

The total axial magnetic field, $B_0 + B_z^s(r)$, is plotted as a function of r in Fig. 3.3.2.[114] Note from this figure that there is 33% field reversal inside the

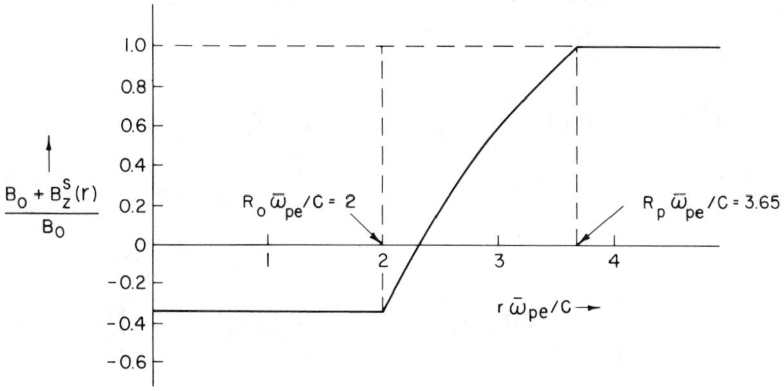

Fig. 3.3.2 Plot of total axial magnetic field, $B_0 + B_z^s(r)$, versus $r\bar{\omega}_{pe}/c$ for the equilibrium distribution fucntion in Eq. (3.3.12) and choice of parameters in Eqs. (3.3.30) and (3.3.31). The axial diamagnetic field $B_z^s(r)$ is defined in Eq. (3.3.20). There is 33% field reversal inside the E-layer $(0 < r < R_0)$.

E-layer $(0 < r < R_0)$ for the choice of parameters in Eqs. (3.3.30) and (3.3.31). Furthermore, the mean azimuthal velocity of the electrons, $V_{e\theta}^0(r)$, is in the *positive* θ-direction,† which is in the same sense in which the electrons would gyrate freely in the uniform external magnetic field $B_0\hat{\mathbf{e}}_z$, for $B_0 > 0$.

3.3.3 E-Layer and Theta-Pinch Equilibrium with Diffuse Radial Boundaries

It should be emphasized that Eq. (3.3.12) is only one of a broad class of electron distribution functions $f_e^0(H, P_\theta)$ that physically correspond to relativistic E-layer equilibria. As an example of an E-layer or linear theta-pinch equilibrium that has diffuse radial boundaries and also permits an exact analytic determination of the axial self magnetic field $B_z^s(r)$, consider the equilibrium distribution function specified by[114]

---

†For example, $V_{e\theta}^0(R_0) = +(\gamma_0^2 - 1)^{1/2}c/\gamma_0$ follows from Eqs. (3.3.17) and (3.3.28) for the choice of parameters in Eqs. (3.3.30) and (3.3.31).

$$f_e^0(H, P_\theta) = \frac{\bar{n}_e}{(2\pi m_e \Theta_e)^{3/2}} \frac{H}{m_e c^2} \exp\left[-(H^2/m_e c^2 - m_e c^2 - 2\omega_e P_\theta)/2\Theta_e\right] , \tag{3.3.32}$$

where H and P_θ are defined in Eqs. (3.3.3) and (3.3.4), ω_e = const., and $\bar{n}_e$ and Θ_e are positive constants. In terms of the phase-space variables (r, $\mathbf{p}$), Eq. (3.3.32) can be expressed in the equivalent form

$$f_e^0(r, \mathbf{p}) = \frac{\bar{n}_e}{(2\pi m_e \Theta_e)^{3/2}} (1 + \mathbf{p}^2/m_e^2 c^2)^{1/2}$$
$$\times \exp\left[-\frac{p_r^2 + (p_\theta - m_e r\omega_e)^2 + p_z^2}{2m_e\Theta_e}\right]$$
$$\times \exp\left\{-\frac{m_e}{2\Theta_e}\left[r^2\omega_e\Omega_e - r^2\omega_e^2 + 2\omega_e e r A_\theta^s(r)/m_e c\right]\right\} , \tag{3.3.33}$$

where use has been made of Eqs. (3.3.3) and (3.3.4). The reason for the somewhat unusual choice of equilibrium distribution function in Eq. (3.3.32) is evident. The inclusion of the multiplicative factor $H/m_e c^2 = (1 + \mathbf{p}^2/m_e^2 c^2)^{1/2}$ in the definition of $f_e^0(H, P_\theta)$ permits the integral $\int d^3p\, (p_\theta/m_e) \times (1 + \mathbf{p}^2/m_e^2c^2)^{-1/2} f_e^0(H, P_\theta)$ to be evaluated in closed form. When Eq. (3.3.32) [or Eq. (3.3.33)] is substituted into Eq. (3.3.8), the azimuthal current density can be expressed as†

$$J_\theta^0(r) = -\bar{n}_e e\omega_e r \exp\left\{-\frac{m_e}{2\Theta_e}\left[r^2\omega_e\Omega_e - r^2\omega_e^2 + 2\omega_e e r A_\theta^s(r)/m_e c\right]\right\} , \tag{3.3.34}$$

and Eq. (3.3.7) reduces to

$$\frac{\partial}{\partial r}\frac{1}{r}\frac{\partial}{\partial r} r A_\theta^s(r) = \frac{4\pi\bar{n}_e e\omega_e r}{c} \exp\left\{-\frac{m_e}{2\Theta_e}\left[r^2\omega_e\Omega_e - r^2\omega_e^2 + 2\omega_e e r A_\theta^s(r)/m_e c\right]\right\} . \tag{3.3.35}$$

Equation (3.3.35) was first solved by Pfirsch.[113] Without loss of generality, the solution to Eq. (3.3.35) that can result in field reversal is

†Note that the expression for $J_\theta^0(r)$ given in Eq. (3.3.34) remains unchanged if the nonrelativistic approximation ($\mathbf{p}^2 \ll m_e^2c^2$) is made in Eqs. (3.3.33) and (3.3.8).

$$rA_\theta^s(r) = -\frac{r^2 B_0}{2} + \frac{c\Theta_e}{e\omega_e}\left[\frac{1}{2}\frac{m_e r^2 \omega_e^2}{\Theta_e} + 2\,\ell n\cosh\left(\frac{r^2 - r_0^2}{2\delta_0^2}\right)\right], \tag{3.3.36}$$

where r_0 is a constant,†

$$\delta_0 \equiv \left(\frac{2c^2\Theta_e/m_e}{\omega_e^2\bar{\omega}_{pe}^2}\right)^{1/4}, \tag{3.3.37}$$

and $\bar{\omega}_{pe}^2 \equiv 4\pi\bar{n}_e e^2/m_e$. When Eq. (3.3.36) is substituted into Eqs. (3.3.11), (3.3.32), and (3.3.34), it is straightforward to show that[114]

$$B_z^s(r) = -B_0 + \frac{c\Theta_e}{e\omega_e}\left[\frac{m_e\omega_e^2}{\Theta_e} + \frac{2}{\delta_0^2}\tanh\left(\frac{r^2 - r_0^2}{2\delta_0^2}\right)\right], \tag{3.3.38}$$

$$f_e^0(r, \mathbf{p}) = \operatorname{sech}^2\left(\frac{r^2 - r_0^2}{2\delta_0^2}\right)\frac{\bar{n}_e}{(2\pi m_e\Theta_e)^{3/2}}(1 + \mathbf{p}^2/m_e^2c^2)^{1/2}$$
$$\times \exp\left\{-\frac{p_r^2 + (p_\theta - m_e r\omega_e)^2 + p_z^2}{2m_e\Theta_e}\right\}, \tag{3.3.39}$$

and

$$J_\theta^0(r) = -\bar{n}_e e r\omega_e \operatorname{sech}^2\left(\frac{r^2 - r_0^2}{2\delta_0^2}\right). \tag{3.3.40}$$

Note from Eqs. (3.3.39) and (3.3.40) that the E-layer has diffuse radial boundaries. Furthermore, from Eq. (3.3.40), the magnitude of the azimuthal current density is a maximum at $r = r_0$, and the radial thickness of the current

---

†The constant r_0 can be related to δ_0, $\bar{n}_e$, Θ_e, and the electron density on axis. Substituting Eq. (3.3.39) into Eq. (3.3.10) gives

$$\operatorname{sech}^2\left(\frac{r_0^2}{2\delta_0^2}\right) = \frac{n_e^0(r=0)}{\bar{n}_e}\left[\frac{1}{(2\pi m_e\Theta_e)^{3/2}}\int d^3p\,(1 + \mathbf{p}^2/m_e^2c^2)^{1/2}\right.$$
$$\left.\times \exp\left(-\frac{p_r^2 + p_\theta^2 + p_z^2}{2m_e\Theta_e}\right)\right]^{-1}.$$

layer is of the order δ_0 [see Eq. (3.3.37)]. Making use of Eq. (3.3.38), we can express the condition that the axial self magnetic field vanishes outside the E-layer $[B_z^s(r \gg r_0) = 0]$ as

$$\frac{eB_0}{m_e c} = \omega_e \left(1 + \frac{2\Theta_e/m_e}{\omega_e^2 \delta_0^2}\right), \tag{3.3.41}$$

which relates B_0, ω_e, Θ_e, and δ_0. For $B_0 > 0$, it follows from Eq. (3.3.41) that $\omega_e > 0$ is required for existence of the equilibrium. This implies that the mean azimuthal velocity of the electrons, $V_{e\theta}^0(r)$, is in the *positive* θ-direction [compare Eqs. (3.3.8) and Eq. (3.3.40)], which is in the same sense in which the electrons would gyrate freely in the uniform external field $B_0\hat{\mathbf{e}}_z$, for $B_0 > 0$. For $\omega_e > 0$, Eq. (3.3.41) can be expressed in the equivalent form

$$\frac{eB_0}{m_e c} = \omega_e + \frac{(2\Theta_e/m_e)^{1/2}\bar{\omega}_{pe}}{c}, \tag{3.3.42}$$

where use has been made of Eq. (3.3.37). When Eq. (3.3.42) is substituted into Eq. (3.3.38), the axial self magnetic field can be expressed as†

$$B_z^s(r) = -\frac{m_e\bar{\omega}_{pe}(2\Theta_e/m_e)^{1/2}}{e}\left[1 - \tanh\left(\frac{r^2 - r_0^2}{2\delta_0^2}\right)\right], \tag{3.3.43}$$

and the self magnetic field on axis reduces to

$$B_z^s(r=0) = -\frac{m_e\bar{\omega}_{pe}(2\Theta_e/m_e)^{1/2}}{e}\left[1 + \tanh\left(\frac{r_0^2}{2\delta_0^2}\right)\right], \tag{3.3.44}$$

If ω_e is close to zero $(0 < \omega_e \ll eB_0/m_e c)$, it follows from Eqs. (3.3.42) and (3.3.44) that

$$B_z^s(r=0) \simeq -B_0\left[1 + \tanh\left(\frac{r_0^2}{2\delta_0^2}\right)\right]. \tag{3.3.45}$$

If $r_0^2 \gg 2\delta_0^2$ is also satisfied, $B_z^s(0) \simeq -2B_0$, and field reversal is almost complete.

As an example, consider the choice of equilibrium parameters related by

---

†Note that the coefficient in Eq. (3.3.43) can also be expressed as

$$m_e\bar{\omega}_{pe}(2\Theta_e/m_e)^{1/2}/e = (8\pi\bar{n}_e\Theta_e)^{1/2}.$$

$$\omega_e = 0.05\, eB_0 \,/\, m_e c \tag{3.3.46}$$

and

$$r_0^2 = 10\,\delta_0^2\ . \tag{3.3.47}$$

Substituting Eqs. (3.3.46) and (3.3.47) into Eqs. (3.3.42) and (3.3.44) gives

$$(2\Theta_e / m_e)^{1/2}\overline{\omega}_{pe} / c = 0.95\, eB_0 / m_e c, \qquad \text{and} \qquad B_z^s(r=0) \simeq -1.9 B_0\,.$$

Furthermore, $\delta_0^2 = 19\, c^2/\overline{\omega}_{pe}^2$ follows from Eq. (3.3.37).

The total axial magnetic field, $B_0 + B_z^s(r)$, is plotted as a function of r in Fig. 3.3.3.[114] Note that there is 90% field reversal for the choice of parameters in Eqs. (3.3.46) and (3.3.47).

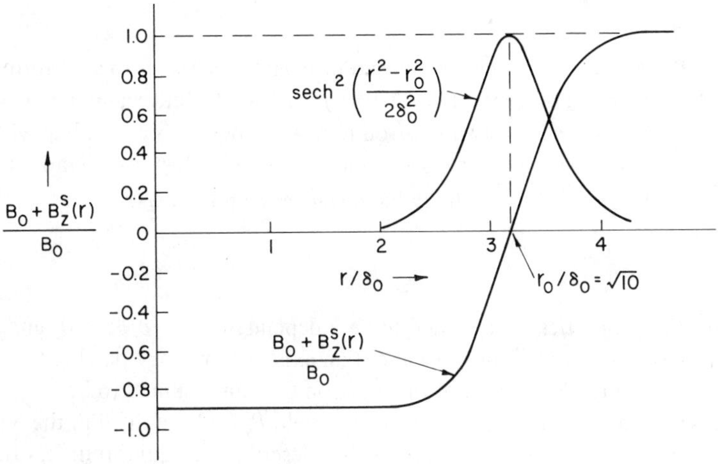

Fig. 3.3.3 Plot of total axial magnetic field, $B_0 + B_z^s(r)$, versus r/δ_0 for the equilibrium distribution function in Eq. (3.3.32) and choice of parameters in Eqs. (3.3.46) and (3.3.47). The axial diamagnetic field $B_z^s(r)$ is defined in Eq. (3.3.43). There is 90% field reversal on axis ($r = 0$).

3.4 RELATIVISTIC ELECTRON BEAM EQUILIBRIA

3.4.1 General Discussion

Solutions to the steady-state ($\partial/\partial t = 0$) Vlasov-Maxwell equations that correspond to intense relativistic electron beam equilibria[122–127] are of considerable practical interest.[32–43] In this section an example of a self-consistent Vlasov equilibrium is considered for a relativistic electron beam propagating parallel to a uniform external magnetic field, $\mathbf{B}_0^{\text{ext}}(\mathbf{x}) = B_0\hat{\mathbf{e}}_z$. Use is made of the steady-state Vlasov-Maxwell description of axisymmetric column equilibria discussed in Section 3.1.2. The equilibrium configuration and cylindrical polar coordinate system (r, θ, z) used in the present analysis are illustrated in Fig. 3.4.1.

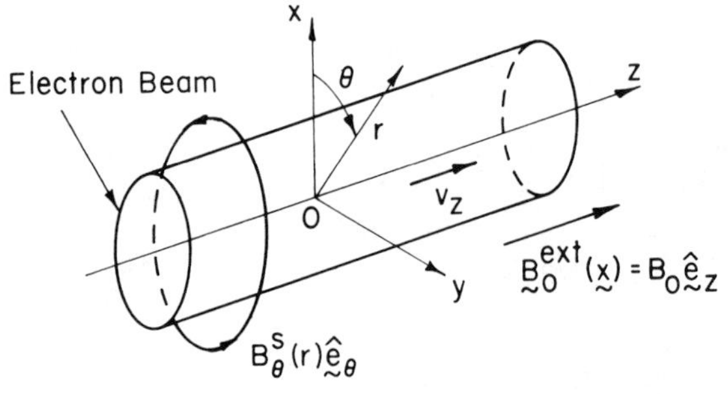

Fig. 3.4.1 Axisymmetric electron beam propagating parallel to a uniform external magnetic field, $\mathbf{B}_0^{\text{ext}}(\mathbf{x}) = B_0\hat{\mathbf{e}}_z$. Cylindrical polar coordinates (r, θ, z) are introduced with the z-axis coinciding with the axis of symmetry; θ is the polar angle in the x-y plane, and $r = \sqrt{x^2 + y^2}$ is the radial distance from the z-axis.

Equilibrium properties are assumed to be independent of z $(\partial/\partial z = 0)$ and azimuthally symmetric $(\partial/\partial\theta = 0)$ about an axis of symmetry parallel to $B_0\hat{\mathbf{e}}_z$. The mean axial motion of the electron beam is assumed to be relativistic, and the azimuthal self magnetic field $B_\theta^s(r)$ is retained in the equilibrium analysis. The relevant equations that describe the equilibrium configuration are Eqs. (3.1.8)–(3.1.15).

To simplify the theoretical analysis the following assumptions are made:

1. The electron beam is immersed in a partially neutralizing ion background with density[†]

$$n_i^0(r) = f n_e^0(r), \tag{3.4.1}$$

[†] In many applications of interest the assumption $n_i^0(r) = f n_e^0(r)$ may be highly idealized. In this regard the equilibrium analysis in Section 3.4 can be extended in a straightforward manner to describe the ion background self-consistently by means of an equilibrium distribution function $f_i^0(\mathbf{x}, \mathbf{p})$ (see the discussion at the end of Section 3.4).

where f = const. = fractional neutralization. It is further assumed that the mean ion velocity is equal to zero in the laboratory frame.

2. External boundaries are sufficiently far removed from the electron beam that their influence on the equilibrium configuration can be ignored.
3. The equilibrium distribution function for the beam electrons does not depend explicitly on the canonical angular momentum P_θ defined in Eq. (3.1.9). In other words, the present analysis is restricted to the class of equilibrium distribution functions of the form

$$f_e^0(\mathbf{x}, \mathbf{p}) = f_e^0(H, P_z), \tag{3.4.2}$$

where H and P_z are the single-particle constants of the motion defined in Eqs. (3.1.8) and (3.1.10). For equilibrium distributions of the form $f_e^0(H, P_z)$, there is no azimuthal rotation of the electron beam, and the axial self magnetic field is equal to zero.

As discussed in Section 3.1.2, any distribution function that is a function only of the single-particle constants of the motion in the equilibrium fields satisfies the steady-state Vlasov equation. For axisymmetric column equilibria, the total energy H and axial canonical momentum P_z of an electron can be expressed as

$$H = (m_e^2 c^4 + c^2 \mathbf{p}^2)^{1/2} - e\phi^0(r), \tag{3.4.3}$$

$$P_z = p_z - \frac{e}{c} A_z^s(r), \tag{3.4.4}$$

where m_e is the electron mass, $-e$ is the electron charge, r is the radial distance from the axis of symmetry, $A_z^s(r)$ is the axial component of the equilibrium vector potential for the azimuthal self magnetic field $B_\theta^s(r)$ [see Eq. (3.1.7)], $\phi^0(r)$ is the equilibrium electrostatic potential, $\mathbf{p}$ is the mechanical momentum, and $\mathbf{p}^2 = p_r^2 + p_\theta^2 + p_z^2$.

The simplifications provided by assumption 3 can be demonstrated as follows. First, note from Eq. (3.4.3) that H is an even function of p_θ. Therefore, $f_e^0(H, P_z)$ is also an even function of p_θ, and the mean azimuthal velocity $V_{e\theta}^0(r)$ of the electron beam is equal to zero, that is,

$$V_{e\theta}^0(r) = \frac{\int d^3p\, v_\theta f_e^0(H, P_z)}{\int d^3p\, f_e^0(H, P_z)} = 0\,, \tag{3.4.5}$$

where $v_\theta = (p_\theta/m_e)(1 + \mathbf{p}^2/m_e^2c^2)^{-1/2}$†. Since the background ion current is

---

†Similarly, $f_e^0(H, P_z)$ is an even function of p_r, and the mean radial velocity is equal to zero,

$$V_{er}^0(r) = \int d^3p\, v_r f_e^0(H, P_z) \Big/ \int d^3p\, f_e^0(H, P_z) = 0.$$

zero (assumption 1), it follows from Eqs. (3.1.6), (3.1.14), and (3.4.5) that no axial self magnetic field is produced by the relativistic electron beam, that is,

$$B_z^s(r) = 0, \tag{3.4.6}$$

for the class of equilibrium distribution functions of the form $f_e^0(H, P_z)$. Furthermore, from Eqs. (3.4.3) and (3.4.4), the equilibrium properties of the electron beam do not depend on the externally applied magnetic field $B_0\hat{\mathbf{e}}_z$ for electron distribution functions of the form $f_e^0(H, P_z)$.

Once the functional form of $f_e^0(H, P_z)$ is specified, the self-field potentials, $\phi^0(r)$ and $A_z^s(r)$, can be determined from the appropriate Maxwell equations, Eqs. (3.1.13) and (3.1.15). Making use of Eq. (3.4.1), we reduce Eq. (3.1.13) to (single ionization is assumed)

$$\frac{1}{r}\frac{\partial}{\partial r} r \frac{\partial}{\partial r}\phi^0(r) = 4\pi e(1-f)\int d^3p\, f_e^0(H, P_z)\,, \tag{3.4.7}$$

where $\int d^3p\, f_e^0(H, P_z)$ is related to the electron density $n_e^0(r)$ by

$$n_e^0(r) = \int d^3p\, f_e^0(H, P_z)\,. \tag{3.4.8}$$

Furthermore, since the background ion current is zero (assumption 1), Eq. (3.1.15) reduces to

$$\frac{1}{r}\frac{\partial}{\partial r} r \frac{\partial}{\partial r} A_z^s(r) = \frac{4\pi e}{c}\int d^3p\, v_z f_e^0(H, P_z)\,, \tag{3.4.9}$$

where $v_z = (p_z/m_e)(1 + \mathbf{p}^2/m_e^2c^2)^{-1/2}$. In Eq. (3.4.9), $-e\int d^3p v_z f_e^0(H,P_z)$ is related to the axial electron current $J_z^0(r)$ by

$$J_z^0(r) = -en_e^0(r)V_{ez}^0(r) = -e\int d^3p\, v_z f_e^0(H, P_z)\,, \tag{3.4.10}$$

where $V_{ez}^0(r)$ is the mean axial velocity of an electron fluid element,

$$V_{ez}^0(r) = \frac{\int d^3p v_z f_e^0(H, P_z)}{\int d^3p\, f_e^0(H, P_z)}\,. \tag{3.4.11}$$

Since H and P_z depend on $\phi^0(r)$ and $A_z^s(r)$, Eqs. (3.4.7) and (3.4.9) are generally nonlinear differential equations for the self-field potentials. Once the functional form of $f_e^0(H, P_z)$ is specified, however, $\phi^0(r)$ and $A_z^s(r)$ can, in principle, be determined from Eqs. (3.4.7) and (3.4.9), and then the self fields obtained from $E_r^0(r) = -\partial\phi^0(r)/\partial r$ and $B_\theta^s(r) = -\partial A_z^s(r)/\partial r$.

3.4.2 Example of a Relativistic Electron Beam Equilibrium

As an example of a relativistic electron beam equilibrium,[122] consider the electron distribution function specified by[†]

$$f_e^0(H, P_z) = \frac{\bar{n}_e}{2\pi\gamma_0 m_e}\,\delta(P_z - \gamma_0 m_e\beta_0 c)\,\delta(H - \gamma_0 m_e c^2)\,, \qquad (3.4.12)$$

where H and P_z are defined in Eqs. (3.4.3) and (3.4.4), β_0 = const., and $\bar{n}_e$ and γ_0 are positive constants. For the equilibrium distribution function defined in Eq. (3.4.12), all of the electrons have the same value of axial canonical momentum ($P_z = \gamma_0 m_e \beta_0 c$) and the same value of total energy ($H = \gamma_0 m_e c^2$). When Eq. (3.4.12) is substituted into Eq. (3.4.8), it is straightforward to show that the density profile for the electron beam is[122]

$$n_e^0(r) = \begin{cases} \bar{n}_e \left[1 + \dfrac{e\phi^0(r)}{\gamma_0 m_e c^2}\right], & 0 < r < R_p\,, \\[2ex] 0, & r > R_p\,, \end{cases} \qquad (3.4.13)$$

where R_p is determined from the extreme of the interval on which the inequality,

$$\left\{ m_e^2 c^4 + c^2 \left[\gamma_0 m_e \beta_0 c + \frac{e}{c} A_z^s(r)\right]^2 \right\}^{1/2} - e\phi^0(r) - \gamma_0 m_e c^2 \leqslant 0\,, \qquad (3.4.14)$$

is satisfied. Note from Eq. (3.4.13) that the electron beam has a sharp radial boundary at $r = R_p$. When Eq. (3.4.13) is substituted into Eq. (3.4.7), Poisson's equation can be expressed as

$$\frac{1}{r}\frac{\partial}{\partial r} r \frac{\partial}{\partial r}\phi^0(r) = \begin{cases} 4\pi e(1-f)\bar{n}_e \left[1 + \dfrac{e\phi^0(r)}{\gamma_0 m_e c^2}\right], & 0 < r < R_p\,, \\[2ex] 0, & r > R_p\,. \end{cases} \qquad (3.4.15)$$

[†] In Eq. (3.4.12) it is *not* assumed that γ_0 and β_0 are related by $\gamma_0^2 = (1 - \beta_0^2)^{-1}$.

Furthermore, substituting Eq. (3.4.12) into Eq. (3.4.10), we obtain for the axial current density[122]

$$J_z^0(r) = \begin{cases} -\dfrac{\bar{n}_e e}{\gamma_0 m_e}\left[\gamma_0 m_e \beta_0 c + \dfrac{e}{c} A_z^s(r)\right], & 0<r<R_p\,, \\ 0, & r>R_p\,, \end{cases} \tag{3.4.16}$$

and Eq. (3.4.9) reduces to

$$\frac{1}{r}\frac{\partial}{\partial r} r \frac{\partial}{\partial r} A_z^s(r) = \begin{cases} \dfrac{4\pi \bar{n}_e e}{\gamma_0 m_e c}\left[\gamma_0 m_e \beta_0 c + \dfrac{e}{c} A_z^s(r)\right], & 0<r<R_p\,, \\ 0, & r>R_p\,. \end{cases} \tag{3.4.17}$$

From Eqs. (3.4.10), (3.4.13), and (3.4.16), the axial velocity profile for the electron beam is

$$V_{ez}^0(r) = \frac{\beta_0 c + (e/\gamma_0 m_e c)\, A_z^s(r)}{1 + e\phi^0(r)/\gamma_0 m_e c^2}\,, \tag{3.4.18}$$

for $0<r<R_p$.

Note from Eq. (3.4.14) that the beam radius R_p cannot be evaluated explicity until $\phi^0(r)$ and $A_z^s(r)$ have been determined from Eqs. (3.4.15) and (3.4.17). Furthermore, the solutions for $\phi^0(r)$ and $A_z^s(r)$ depend on R_p. Therefore the condition that determines R_p is, in effect, nonlinear. The solutions to Eqs. (3.4.15) and (3.4.17) that are continuous at $r = R_p$, have continuous first derivatives at $r = R_p$, and satisfy $A_z^s(r=0) = 0 = \phi^0(r=0)$ are

$$\phi^0(r) = \begin{cases} -\dfrac{\gamma_0 m_e c^2}{e}\left\{1 - I_0\left[r(1-f)^{1/2}/\delta\right]\right\}, & 0<r<R_p\,, \\ -\dfrac{\gamma_0 m_e c^2}{e}\left\{1 - I_0\left[R_p(1-f)^{1/2}/\delta\right] \right. \\ \left. \quad -\dfrac{R_p}{\delta}(1-f)^{1/2} I_1\left[R_p(1-f)^{1/2}/\delta\right] \ln \dfrac{r}{R_p}\right\}, & r>R_p\,, \end{cases} \tag{3.4.19}$$

$$A_z^s(r) = \begin{cases} -\dfrac{\gamma_0 m_e c^2 \beta_0}{e}\left[1 - I_0(r/\delta)\right], & 0<r<R_p, \\ -\dfrac{\gamma_0 m_e c^2 \beta_0}{e}\left[1 - I_0(R_p/\delta) - \dfrac{R_p}{\delta} I_1(R_p/\delta)\,\ln\dfrac{r}{R_p}\right], & r>R_p. \end{cases} \tag{3.4.20}$$

In Eqs. (3.4.19) and (3.4.20) I_n is the modified Bessel function of order n, and δ is the collisionless skin depth, defined by†

$$\delta^2 = \frac{\gamma_0 c^2}{\overline{\omega}_{pe}^2}, \tag{3.4.21}$$

where $\overline{\omega}_{pe}^2 = 4\pi\overline{n}_e e^2/m_e$.

From Eqs. (3.4.19) and (3.4.20), the radial electric field, $E_r^0(r) = -\partial\phi^0(r)/\partial r$, and the azimuthal self magnetic field, $B_\theta^s(r) = -\partial A_z^s(r)/\partial r$, can be expressed as

$$E_r^0(r) = \begin{cases} -\dfrac{\gamma_0 m_e c^2}{e}\dfrac{(1-f)^{1/2}}{\delta} I_1[r(1-f)^{1/2}/\delta], & 0<r<R_p, \\ -\dfrac{\gamma_0 m_e c^2}{e}\dfrac{R_p(1-f)^{1/2}}{r\delta} I_1[R_p(1-f)^{1/2}/\delta], & r>R_p, \end{cases} \tag{3.4.22}$$

and

$$B_\theta^s(r) = \begin{cases} -\dfrac{\gamma_0 m_e c^2 \beta_0}{e\delta} I_1(r/\delta), & 0<r<R_p, \\ -\dfrac{\gamma_0 m_e c^2 \beta_0}{e\delta}\dfrac{R_p}{r} I_1(R_p/\delta), & r>R_p. \end{cases} \tag{3.4.23}$$

†This definition of δ should not be confused with Eq. (3.3.19).

Furthermore, when Eqs. (3.4.19) and (3.4.20) are combined with Eqs. (3.4.13) and (3.4.18), the electron density profile and axial velocity profile are

$$n_e^0(r) = \begin{cases} \bar{n}_e I_0\,[r(1-f)^{1/2}/\delta]\,, & 0<r<R_p\,, \\ 0, & r>R_p\,, \end{cases} \tag{3.4.24}$$

and

$$V_{ez}^0(r) = \beta_0 c\,\frac{I_0(r/\delta)}{I_0\,[r(1-f)^{1/2}/\delta]}\,, \tag{3.4.25}$$

for $0<r<R_p$. Since $I_0(0)=1$, it follows from Eqs. (3.4.24) and (3.4.25) that $\bar{n}_e =$ const. and $\beta_0 c =$ const. can be identified with the values of electron density and mean axial velocity at $r=0$, respectively.

Since the equilibrium distribution function in Eq. (3.4.12) selects $H=\gamma_0 m_e c^2$, the mechanical energy of an electron at radius r is $\gamma_e^0(r) m_e c^2 = (m_e^2 c^4 + c^2 p_\perp^2 + c^2 p_z^2)^{1/2} = e\phi^0(r) + \gamma_0 m_e c^2$, where $p_\perp = (p_r^2 + p_\theta^2)^{1/2}$ is the electron momentum transverse to the beam. Making use of Eq. (3.4.19), we can express $\gamma_e^0(r)$ as

$$\gamma_e^0(r) = \gamma_0 I_0\,[r(1-f)^{1/2}/\delta] \tag{3.4.26}$$

for $0<r<R_p$. Note that $\gamma_e^0(r)$ has the same functional form as the electron density profile $n_e^0(r)$ [Eq. (3.4.24)], and that $\gamma_0 m_e c^2 =$ const. can be identified with the energy of an electron as it passes through the axis of the beam ($r=0$).

It is also straightforward to calculate the total current carried by the electron beam for the equilibrium distribution function defined in Eq. (3.4.12). Assuming $\beta_0>0$ and making use of Eqs. (3.4.24) and (3.4.25), we can express the *magnitude* of the total beam current as

$$I = 2\pi e\int_0^{R_p} dr\, r n_e^0(r) V_{ez}^0(r) = (m_e c^3/e)\,(R_p/2\delta)\beta_0\gamma_0 I_1(R_p/\delta). \tag{3.4.27}$$

Evidently, I is sensitive to the value of R_p/δ. For a completely neutralized electron beam ($f=1$) it is instructive to express the right-hand side of Eq. (3.4.27) in units of the Alfvén limiting current I_A.[142] On the basis of Lawson's definition of I_A for a nonuniform beam, I_A is the beam current at which the Larmor radius of an electron in the maximum self magnetic field (at $r=R_p$) is equal to one-half of the beam radius:[94–96]

$$\frac{\gamma_e^0(R_p) m_e c V_{ez}^0(R_p)}{e B_\theta^s(R_p)} = \frac{R_p}{2} . \tag{3.4.28}$$

Making use of Eqs. (3.4.24)-(3.4.28), we can express I_A as

$$I_A = (m_e c^3 / e) \beta_0 \gamma_0 I_0(R_p / \delta) \simeq 17{,}000 \, \beta_0 \gamma_0 I_0(R_p / \delta) \text{ amperes.} \tag{3.4.29}$$

Equation (3.4.27) then becomes

$$I = I_A \frac{R_p}{2\delta} \frac{I_1(R_p/\delta)}{I_0(R_p/\delta)} . \tag{3.4.30}$$

Making use of the asymptotic expansions

$$I_n(\lambda) \simeq \frac{\exp(\lambda)}{(2\pi\lambda)^{1/2}} \left(1 - \frac{4n^2 - 1}{8\lambda} + \cdots\right), \quad \text{for } \lambda \gg 1, \tag{3.4.31}$$

and

$$I_n(\lambda) \simeq \frac{1}{n!} \left(\frac{\lambda}{2}\right)^n \left[1 + \frac{\lambda^2}{4(n+1)} + \cdots\right], \quad \text{for } |\lambda| \ll 1, \tag{3.4.32}$$

it follows from Eq. (3.4.30) that

$$I \simeq \frac{R_p}{2\delta} I_A \gg I_A, \quad \text{for } R_p \gg \delta , \tag{3.4.33}$$

and

$$I \simeq \frac{R_p^2}{4\delta^2} I_A \ll I_A, \text{ for } R_p \ll \delta. \tag{3.4.34}$$

Equation (3.4.33) states that the beam current I exceeds the Alfvén current I_A by a large amount whenever the beam radius is much larger than the collisionless skin depth, $R_p \gg \delta$. However, if $R_p \ll \delta$, the beam current is much less than the Alfvén current [Eq. (3.4.34)].

The ratio of beam radius to collisionless skin depth, R_p/δ, can be related to ν/γ_0, where $\nu = \bar{N}_e r_0$ is Budker's parameter,[30] $\bar{N}_e$ is the total number of electrons per unit length of the beam, and $r_0 = e^2/m_ec^2$ is the classical electron radius. Making use of Eq. (3.4.24), it is straightforward to show that

$$\overline{N}_e = 2\pi \int_0^{R_p} dr\, r n_e^0(r) = (\gamma_0 m_e c^2/e^2)(1-f)^{-1/2}(R_p/2\delta) I_1 [R_p(1-f)^{1/2}/\delta] .$$

Therefore ν/γ_0 can be expressed as

$$\frac{\nu}{\gamma_0} = \frac{1}{2(1-f)^{1/2}} \frac{R_p}{\delta} I_1 [R_p(1-f)^{1/2}/\delta] . \qquad (3.4.35)$$

If the beam is completely neutralized ($f = 1$), Eq. (3.4.35) reduces to [see Eq. (3.4.32)]

$$\frac{\nu}{\gamma_0} = \frac{R_p^2}{4\delta^2} , \qquad (3.4.36)$$

Evidently, $R_p \lesseqgtr 2\delta$ accordingly as $\nu/\gamma_0 \lesseqgtr 1$. It follows from Eqs. (3.4.33) and (3.4.36) that $\nu/\gamma_0 \gg 1$ is required for the beam current I to exceed the Alfvén current I_A by a large amount.

The beam radius R_p is determined from the extreme of the interval on which the inequality in Eq. (3.4.14) is satisfied. When Eqs. (3.4.19) and (3.4.20) are substituted into Eq. (3.4.14), R_p is the solution to[122]

$$1 = \gamma_0^2 \{ I_0^2 [R_p(1-f)^{1/2}/\delta] - \beta_0^2 I_0^2 (R_p/\delta) \} . \qquad (3.4.37)$$

Note that Eq. (3.4.37) determines R_p in terms of f and parameters that characterize the equilibrium distribution function defined in Eq. (3.4.12) (e.g., β_0, γ_0, and $\overline{n}_e$). Once the values of f, β_0, and γ_0 are specified, R_p/δ can be evaluated numerically from Eq. (3.4.37), where δ is defined in Eq. (3.4.21). Since R_p/δ is required to be real, there are restrictions on the allowed values of f, β_0^2, and γ_0^2 in Eq. (3.4.37). Keep in mind that $\gamma_0 m_e c^2$ and $\beta_0 c$ are the energy and axial velocity, respectively, of an electron at $r = 0$; the transverse speed $\beta_\perp c$ of an electron as it passes through the axis ($r = 0$) of the beam is related to γ_0 and β_0 by

$$\frac{1}{1 - \beta_\perp^2 - \beta_0^2} = \gamma_0^2 . \qquad (3.4.38)$$

Therefore Eq. (3.4.37) can be expressed in the equivalent form

$$\beta_\perp^2 = \beta_0^2 [I_0^2(R_p/\delta) - 1] - \{ I_0^2 [R_p(1-f)^{1/2}/\delta] - 1 \} . \qquad (3.4.39)$$

Making use of Eq. (3.4.39), we can express the conditions $\beta_\perp^2 > 0$ and $\beta_\perp^2 + \beta_0^2 < 1$ as

$$\frac{I_0^2[R_p(1-f)^{1/2}/\delta] - 1}{I_0^2(R_p/\delta) - 1} < \beta_0^2 < \frac{I_0^2[R_p(1-f)^{1/2}/\delta]}{I_0^2(R_p/\delta)}, \tag{3.4.40}$$

which places restrictions on the values of β_0 and f in order for Eq. (3.4.39) [or Eq. (3.4.37)] to have real solutions for R_p/δ (and hence for the equilibrium to exist). For example, since $\beta_0^2 < 1$ is required, it follows from Eq. (3.4.40) that the equilibrium does not exist if the beam is completely unneutralized ($f = 0$). For $R_p/\delta \ll 1$, note that the first inequality in Eq. (3.4.40) reduces to [see Eq. (3.4.32)]

$$\beta_0^2 > 1 - f, \tag{3.4.41}$$

which is the familiar condition that magnetic pinching forces exceed electrostatic repulsive forces for radial confinement of an electron beam with uniform density and current density (see Section 2.4). For $R_p/\delta \gg 1$ and practical values of $\gamma_e^0(R_p)$, f is required to be very close to unity. For example, if $R_p/\delta = 24.5$, $\gamma_0 = 2$, and $f = 0.98$, then $\gamma_e^0(R_p) \simeq 12$ follows from Eq. (3.4.26). However, if $R_p/\delta = 24.5$, $\gamma_0 = 2$, and $f = 0.5$, then $\gamma_e^0(R_p) \simeq 4 \times 10^6$!

It is evident from Eqs. (3.4.22)-(3.4.27) and Eq. (3.4.35) that the equilibrium properties of the electron beam are quite different in the two regimes: (*a*) $R_p \ll \delta$, and (*b*) $R_p \gg \delta$. In circumstances where the beam radius is much smaller than the collisionless skin depth, $R_p \ll \delta$, it follows from Eqs. (3.4.24)-(3.4.26) and Eq. (3.4.32) that $n_e^0(r) \simeq \bar{n}_e$, $V_{ez}^0(r) \simeq \beta_0 c$, and $\gamma_e^0(r) \simeq \gamma_0$, for $0 < r < R_p$. In other words, the electron density, axial velocity, and energy profiles are approximately uniform within the beam. Furthermore, from Eqs. (3.4.22), (3.4.23) and (3.4.32) the magnitudes of the radial electric field, $E_r^0(r)$, and azimuthal self magnetic field, $B_\theta^s(r)$, increase *linearly* with r from $r = 0$ to $r = R_p$. Making use of Eqs. (3.4.32) and (3.4.39), we can express the beam radius, when $R_p \ll \delta$, in the approximate form

$$R_p^2 = \frac{2\beta_\perp^2}{\beta_0^2 - (1-f)}\delta^2, \tag{3.4.42}$$

which determines R_p in terms of f, β_0^2, δ, and $\beta_\perp^2 (= 1 - \beta_0^2 - \gamma_0^{-2})$. Note that $2\beta_\perp^2 \ll \beta_0^2 - (1-f)$ is required for $R_p \ll \delta$ to be consistent. For the equilibrium distribution function defined in Eq. (3.4.12), it is evident that the equilibrium properties of the electron beam are simple to calculate when $R_p \ll \delta$. However, this regime is of little practical interest from the standpoint of high-current relativistic electron beams.[122] From Eqs. (3.4.27) and (3.4.32) the total current carried by the electron beam for $R_p \ll \delta$ is $I \simeq (mc^3/e)\beta_0\gamma_0(R_p^2/4\delta^2) \ll 17{,}000\,\beta_0\gamma_0$ amperes [see Eq. (3.4.34)].

In circumstances where the beam radius is much larger than the collisionless skin depth, $R_p \gg \delta$, the equilibrium properties of the electron beam are highly nonuniform [see Eqs. (3.4.22)-(3.4.26)]. For example, when $f \neq 1$, it follows from Eqs. (3.4.24)-(3.4.26) that $n_e^0(r)$, $V_{ez}^0(r)$, and $\gamma_e(r)$ increase monotonically† as functions of r from $r=0$ to $r=R_p$. The equilibrium properties simplify somewhat when the beam is completely neutralized ($f=1$). In this case the equilibrium radial electric field is equal to zero [Eq. (3.4.22)], and the energy profile is uniform across the beam, that is, $\gamma_e(r)=\gamma_0$ for $0<r<R_p$ [Eq. (3.4.26)].

Furthermore, for $f=1$, it follows from Eqs. (3.4.24) and (3.4.25) that the beam density is constant,

$$n_e^0(r) = \begin{cases} \bar{n}_e\,, & 0<r<R_p\,, \\ 0, & r>R_p\,, \end{cases} \tag{3.4.43}$$

and the axial velocity profile can be expressed as

$$V_{ez}^0(r) = \beta_0 c I_0(r/\delta) \tag{3.4.44}$$

for $0<r<R_p$. Evidently, $V_{ez}^0(r)$ increases monotonically from the value $\beta_0 c$ (at $r=0$) to the value

$$V_{ez}^0(R_p) = \beta_0 c I_0(R_p/\delta) \tag{3.4.45}$$

at $r=R_p$. If the beam radius is much larger than the collisionless skin depth, $R_p \gg \delta$, it follows from Eq. (3.4.45) that $V_{ez}^{02}(R_p)/c^2 \equiv \beta_{ez}^2(R_p) \gg \beta_0^2$. Since $\beta_{ez}^2(R_p)<1$ is required, $R_p \gg \delta$ is consistent only if $\beta_0^2 \ll 1$.‡ Therefore, for $R_p/\delta \gg 1$, the axial current density $-en_e^0(r)V_{ez}^0(r)$ is highly nonuniform and

---

†Keep in mind that $I_n(x)$ increases monotonically as x increases [see Eqs. (3.4.31) and (3.4.32)].

---

‡For $R_p/\delta \gg 1$, the electron motion near the axis ($r=0$) is predominately in the transverse direction. For $f=1$, Eq. (3.4.39) can be expressed as

$$\beta_\perp^2 = \beta_0^2\,[I_0^2(R_p/\delta) - 1]$$

where $\beta_\perp c$ is the transverse speed of an electron as it passes through $r=0$ [see Eq. (3.4.38)]. For $R_p \gg \delta$, it follows that $\beta_\perp^2 \gg \beta_0^2$.

most of the current is carried in a thin layer (thickness $\approx \delta$) near the surface of the electron beam. For $f = 1$, Eqs. (3.4.37) and (3.4.39) reduce to

$$I_0^2(R_p / \delta) = \frac{\gamma_0^2 - 1}{\gamma_0^2 \beta_0^2} = \frac{\beta_\perp^2 + \beta_0^2}{\beta_0^2}, \tag{3.4.46}$$

which determines the beam radius R_p in terms of δ, β_0, and γ_0 (or $\beta_\perp$). Evidently, for $R_p \gg \delta$ to be satisfied requires $\gamma_0^2 - 1 \gg \gamma_0^2\beta_0^2$. As an example, when $(\gamma_0^2 - 1)/\gamma_0^2\beta_0^2 = 127.73$, Eq. (3.4.46) gives $R_p \simeq 4.00\,\delta$. The corresponding radial profiles for $n_e^0(r)$ [Eq. (3.4.43)], $V_{ez}^0(r)$ [Eq. (3.4.44)], and $B_\theta^s(r)$ [Eq. (3.4.23)] are plotted in Figs. 3.4.2-3.4.4. Note from Fig. 3.4.3 that the axial velocity profile is strongly peaked at $r = R_p$ with $V_{ez}^0(R_p) = 11.30\,\beta_0 c$. Since most of the current is carried in a thin layer near the surface of the electron beam, the magnitude of the self magnetic field is also strongly peaked at $r = R_p$(Fig. 3.4.4). For the example illustrated in Figs. 3.4.2-3.4.4, $\nu/\gamma_0 = 4$ [Eq. (3.4.36)], and the total current carried by the electron beam is $I = 1.73\,I_A$ [Eq. (3.4.30)].

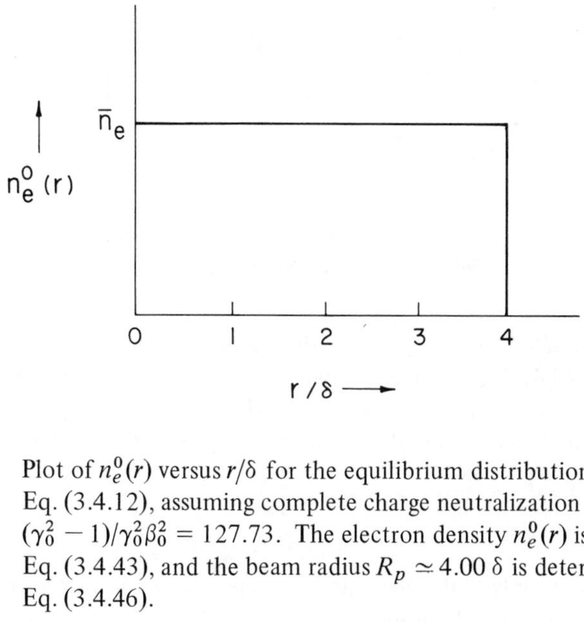

Fig. 3.4.2 Plot of $n_e^0(r)$ versus r/δ for the equilibrium distribution function in Eq. (3.4.12), assuming complete charge neutralization ($f = 1$) and $(\gamma_0^2 - 1)/\gamma_0^2\beta_0^2 = 127.73$. The electron density $n_e^0(r)$ is defined in Eq. (3.4.43), and the beam radius $R_p \simeq 4.00\,\delta$ is determined from Eq. (3.4.46).

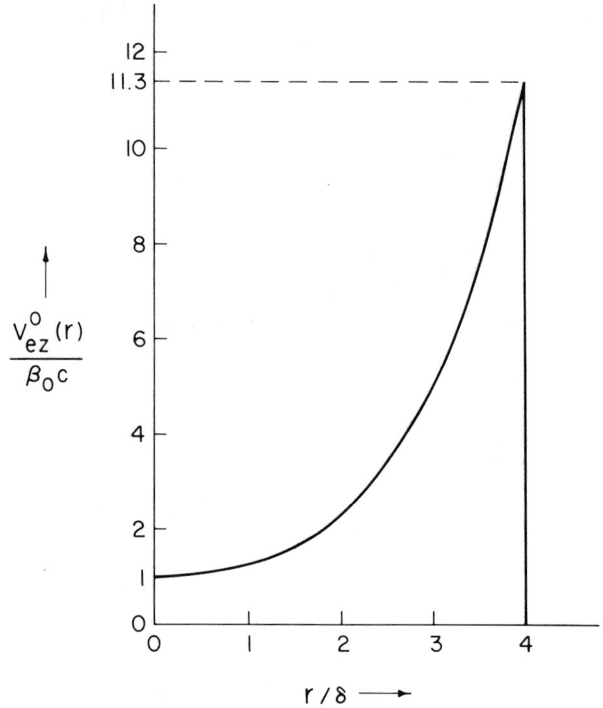

Fig. 3.4.3 Plot of $V^0_{ez}(r)$ versus r/δ for the equilibrium distribution function in Eq. (3.4.12), assuming complete charge neutralization ($f = 1$) and $(\gamma_0^2 - 1)/\gamma_0^2\beta_0^2 = 127.73$. The mean axial velocity $V^0_{ez}(r)$ is defined in Eq. (3.4.44), and the beam radius $R_p \simeq 4.00\ \delta$ is determined from Eq. (3.4.46).

It is important to note that other equilibrium properties of the electron beam can also be calculated for the equilibrium distribution function defined in Eq. (3.4.12). For example, it can be shown from Eqs. (1.3.22) and (3.4.12) that the equilibrium pressure tensor for the electrons is of the form

$$\boldsymbol{P}^0_e(\mathbf{x}) = n^0_e(r)T^0_{e\perp}(r)\ (\hat{\mathbf{e}}_r\hat{\mathbf{e}}_r + \hat{\mathbf{e}}_\theta\hat{\mathbf{e}}_\theta), \tag{3.4.47}$$

where $\hat{\mathbf{e}}_r$ and $\hat{\mathbf{e}}_\theta$ are unit vectors in the r- and θ-directions, respectively, and the transverse electron temperature $T^0_{e\perp}(r)$ is defined by

$$T^0_{e\perp}(r) = \frac{1}{2m_e}\ \frac{\displaystyle\int d^3p\ \frac{p_r^2 + p_\theta^2}{(1 + \mathbf{p}^2/m_e^2c^2)^{1/2}} f^0_e(H, P_z)}{\displaystyle\int d^3p\, f^0_e(H, P_z)}. \tag{3.4.48}$$

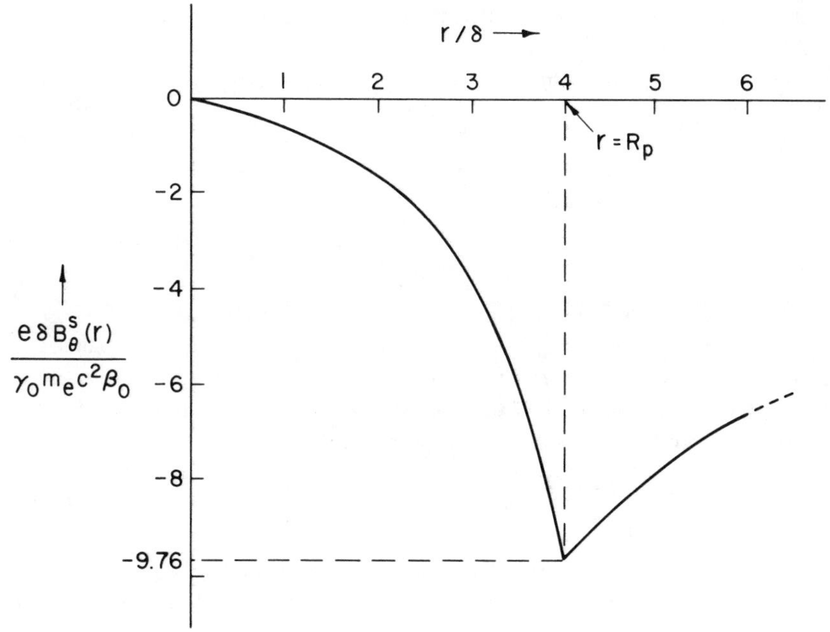

Fig. 3.4.4 Plot of $B_\theta^s(r)$ versus r/δ for the equilibrium distribution in Eq. (3.4.12), assuming complete charge neutralization ($f = 1$) and $(\gamma_0^2 - 1)/\gamma_0^2\beta_0^2 = 127.73$. The azimuthal self magnetic field is defined in Eq. (3.4.23), and the beam radius $R_p \simeq 4.00\ \delta$ is determined from Eq. (3.4.46).

Equation (3.4.48) follows since $f_e^0(H, P_z)$ is an even function of p_r and p_θ [see Eq. (3.4.3)] and is symmetric under interchange of p_r^2 and p_θ^2. Substituting Eq. (3.4.12) into Eq. (3.4.48), and making use of Eqs. (3.4.19) and (3.4.20), it is straightforward to show that

$$T_{e\perp}^0(r) = \frac{m_e c^2/2}{\gamma_0 I_0\,[r(1-f)^{1/2}/\delta]}\left\{\gamma_0^2 I_0^2\,[r(1-f)^{1/2}/\delta] - \gamma_0^2\beta_0^2 I_0^2(r/\delta) - 1\right\} \tag{3.4.49}$$

for $0 < r < R_p$. The transverse electron temperature decreases monotonically from $T_{e\perp}^0(r=0) = \gamma_0 m_e c^2 \beta_\perp^2/2$, on axis, to $T_{e\perp}^0(r = R_p) = 0$, at the edge of the beam [see Eqs. (3.4.37) and (3.4.38)]. Note from Eq. (3.4.47) that the electrons are *cold* in the z-direction since $[P_e^0(\mathbf{x})]_{zz} = 0$ for the choice of distribution function in Eq. (3.4.12).

It should be emphasized that Eq. (3.4.12) is only one of a broad class of electron distribution functions $f_e^0(H, P_z)$ that physically correspond to relativistic electron beam equilibria.[122-127] Therefore the choice of $f_e^0(H, P_z)$ should be guided by what seems appropriate in a given experimental situation. For example, to incorporate a thermal spread in axial velocities, a possible choice of $f_e^0(H, P_z)$ is

$$f_e^0(H, P_z) = N_0 \exp\left\{-\frac{(P_z - \gamma_0 m_e \beta_0 c)^2}{2m_e \Theta_{ez}}\right\} \delta(H - \gamma_0 m_e c^2)\,, \tag{3.4.50}$$

where $\beta_0 =$ const., and N_0, Θ_{ez}, and γ_0 are positive constants.

In concluding this section, it is important to note that assumption 3 can be relaxed in a straightforward manner to describe relativistic electron beam equilibria with mean azimuthal rotation. In this case, the equilibrium distribution function for the electrons is of the form [see Eq. (3.1.12)]

$$f_e^0(\mathbf{x}, \mathbf{p}) = f_e^0(H, P_\theta, P_z)\,, \tag{3.4.51}$$

where H and P_z are defined in Eqs. (3.4.3) and (3.4.4), and P_θ is the canonical angular momentum, defined in Eq. (3.1.9). For an electron, P_θ can be expressed as

$$P_\theta = r\left[p_\theta - \frac{e}{c}A_\theta^{\text{ext}}(r) - \frac{e}{c}A_\theta^s(r)\right], \tag{3.4.52}$$

where $A_\theta^{\text{ext}}(r) = rB_0/2$ (for a uniform external guide field $B_0\hat{\mathbf{e}}_z$), and $A_\theta^s(r)$ is the θ-component of vector potential for the axial self magnetic field, $B_z^s(r) = \partial A_\theta^s(r)/\partial r + A_\theta^s(r)/r$, generated by the azimuthal motion of the electron beam. Since the background ion current is zero (assumption 1), Eq. (3.1.14) can be expressed as

$$\frac{\partial}{\partial r}\frac{1}{r}\frac{\partial}{\partial r} rA_\theta^s(r) = \frac{4\pi e}{c}\int d^3p\, v_\theta f_e^0(H, P_\theta, P_z)\,, \tag{3.4.53}$$

which determines $A_\theta^s(r)$ [and hence $B_z^s(r)$] self-consistently in terms of $f_e^0(H, P_\theta, P_z)$. The self field potentials, $\phi^0(r)$ and $A_z^s(r)$, are of course determined from Eqs. (3.4.7) and (3.4.9) with $f_e^0(H, P_z)$ replaced by $f_e^0(H, P_\theta, P_z)$.

As an example of a rotating relativistic electron beam equilibrium, consider the electron distribution function specified by[143]

$$f_e^0(H, P_\theta, P_z) = N_0\, \delta(P_z - \gamma_0 m_e \beta_0 c)\, \delta(H - \omega_e P_\theta - \gamma_0 m_e c^2)\,, \tag{3.4.54}$$

where $\beta_0 =$ const., $\omega_e =$ const., and N_0 and γ_0 are positive constants. For $\omega_e = 0$, Eq. (3.4.54) reduces to Eq. (3.4.12) (with $N_0 \equiv \hat{n}_e/2\pi\gamma_0 m_e$), and there is no rotation of the electron beam. For $\omega_e \neq 0$, however, the mean azimuthal velocity $V_{e\theta}^0(r)$ is nonzero. For small values of ω_e it can be shown that the mean azimuthal motion of the electron beam corresponds to a rigid rotation with angular velocity $V_{e\theta}^0(r)/r = \omega_e$.

It should also be noted that assumption 1 $[n_i^0(r) = fn_e^0(r)]$ is highly idealized and probably is not satisfied in many applications of interest. In this regard, it is straightforward to extend the equilibrium analysis in Section 3.4 to describe the positive ion background within the framework of the steady-state Vlasov-Maxwell equations. If the ions are described by an equilibrium distribution function $f_i^0(H)$,† the procedure for calculating the equilibrium properties of the electron beam from $f_e^0(H, P_\theta, P_z)$ remains essentially the same. The only difference is that the electrostatic potential $\phi^0(r)$ must be calculated self-consistently using the ion density computed from $f_i^0(H)$, that is, the right-hand side of Eq. (3.4.7) must be replaced by

$$-4\pi\rho^0(r) = 4\pi e\left[\int d^3p\, f_e^0(H, P_\theta, P_z) - n_i^0(r)\right] \tag{3.4.55}$$

where $n_i^0(r) \equiv \int d^3p\, f_i^0(H)$.

3.5 RELATIVISTIC ELECTRON RING EQUILIBRIA

3.5.1 General Discussion

The properties of ring currents of relativistic electrons, contained in a magnetic mirror field, have been extensively studied in recent years in connection with electron ring particle accelerators.[10–27] Theoretical studies of the equilibrium and stability properties of relativistic electron rings have been undertaken from two different points of view. The traditional approach is based on single-particle orbit calculations for rather simple equilibrium configurations, making use of the concept of betatron oscillations about an equilibrium orbit. This approach lacks both the flexibility and generality required for investigating more complex equilibria and the collective instabilities that may develop. The Vlasov-Maxwell equations, however, provide a natural framework for an analysis of the equilibrium and stability properties of collisionless charged particle systems (electrically neutral or nonneutral) in which collective effects are important.

---

†Note from Eq. (3.1.30) that H is an even function of p_r, p_θ, and p_z. Therefore the mean ion velocity is $\mathbf{V}_i^0(\mathbf{x}) = \int d^3p\, \mathbf{v} f_i^0(H) = 0$ (see assumption 1). If the ion dynamics are nonrelativistic, $H = \mathbf{p}^2/2m_i + e\phi^0(r) + m_i c^2$, where m_i and $+e$ are the ion mass and charge, respectively.

In this section, an example of a relativisitic electron ring equilibrium is considered which incorporates a thermal spread in energy H of the electrons.[132] Use is made of the steady-state ($\partial/\partial t = 0$) Vlasov-Maxwell description of axisymmetric mirror-confined equilibria discussed in Section 3.1.3 [see Eqs. (3.1.30)-(3.1.34) and Fig. 3.1.2]. Equilibria appropriate to electron ring accelerators [25,129-132] possess the essential simplifications that the ratio of minor dimensions to major dimensions of the ring is small, and that the transverse velocity is small in comparison with the component of velocity in the direction of the orbit. Therefore the transverse and longitudinal phase spaces effectively decouple. Although collective effects make very little difference to the major radius of the ring in the regime of experimental interest, they can be important in determining the minor dimensions of the ring (especially in fields with little or no intrinsic axial focusing), and also the stability properties of the ring.

The equilibrium configuration is illustrated in Fig. 3.5.1. It consists of a relativistic electron ring located at the midplane of an externally imposed mirror field $\mathbf{B}_0^{ext}(\mathbf{x})$. The mirror field acts to confine the ring both axially and radially. The equilibrium radius of the ring is denoted by R_0, and the minor dimensions of the ring are designated as $2a$ (radial dimension) and $2b$ (axial dimension). As shown in Fig. 3.5.1, a cylindrical polar coordinate system (r, θ, z) is introduced with the z-axis along the axis of symmetry and $z = 0$ coinciding with the midplane; r is the radial distance from the z-axis, and θ is the polar angle in a plane perpendicular to the z-axis. The equilibrium properties are assumed to be azimuthally symmetric ($\partial/\partial\theta = 0$) about the z-axis. The electrons composing the ring gyrate in the external field $\mathbf{B}_0^{ext}(\mathbf{x})$ with azimuthal velocities v_θ in the positive θ-direction. The associated ring current, which is in the negative θ-direction, produces a self magnetic field $\mathbf{B}_0^s(\mathbf{x})$ which threads the ring in the sense indicated in Fig. 3.5.1. This self magnetic field acts as a *focusing* field which tends to compress the minor dimensions of the ring both axially and radially. The electron ring is assumed to be partially neutralized by a positive ion background. The excess electrons form a potential well for the ions. The electrostatic forces on the electrons, however, are repulsive, that is, the self electric field $\mathbf{E}^0(\mathbf{x})$ acts as a *defocusing* field which tends to increase the minor dimensions of the ring. Although the external magnetic field $\mathbf{B}_0^{ext}(\mathbf{x})$ is depicted as a mirror field in Fig. 3.5.1, it should be pointed out that the present analysis is also applicable if $\mathbf{B}_0^{ext}(\mathbf{x})$ corresponds to a uniform axial field with $\mathbf{B}_0^{ext}(\mathbf{x}) = B_0\hat{\mathbf{e}}_z$, provided the self magnetic field generated by the ring current is sufficiently strong to confine the ring axially, and the assumptions enumerated below are not violated.

To make the theoretical analysis tractable, the following simplifying assumptions are made:

1. The positive ions form a partially neutralizing background. The equilibrium

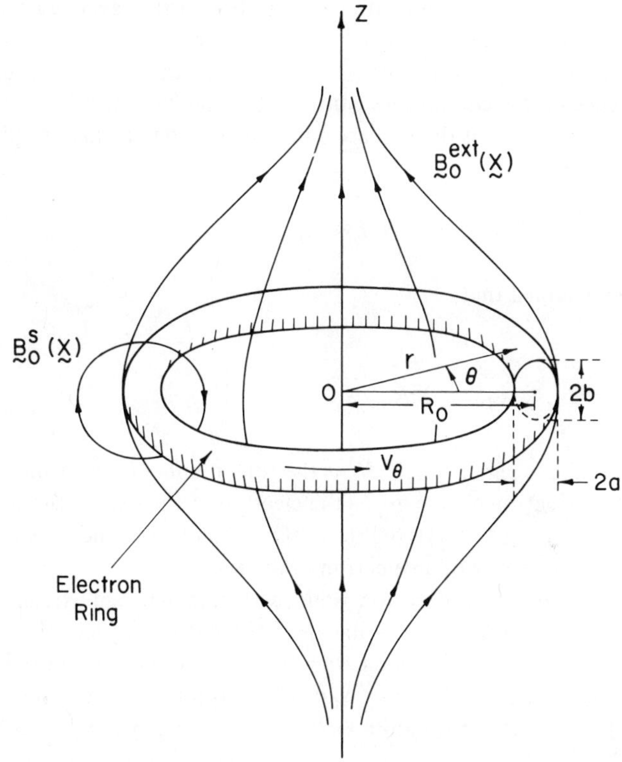

Fig. 3.5.1 Axisymmetric equilibrium configuration for a relativistic electron ring confined in an external mirror field $\mathbf{B}_0^{ext}(\mathbf{x})$. Cylindrical polar coordinates (r, θ, z) are introduced with the z-axis coinciding with the axis of symmetry, and $z = 0$ at the mirror midplane; r is the radial distance from the axis of symmetry, and θ is the polar angle.

ion density $n_i^0(r, z)$ and electron density $n_e^0(r, z)$ are assumed to be related by†

$$n_i^0(r, z) = fn_e^0(r, z) , \qquad (3.5.1)$$

†In many applications of interest the assumption $n_i^0(r, z) = fn_e^0(r, z)$ may be highly idealized. In this regard the equilibrium analysis in Section 3.5 can be extended in a straightforward manner to describe the ion background self-consistently by means of an equilibrium distribution function $f_i^0(\mathbf{x}, \mathbf{p})$ (see the discussion at the end of Section 3.5).

where $f=$ const. $=$ fractional neutralization. It is further assumed that the ion current is equal to zero in the laboratory frame.

2. External boundaries are sufficiently far removed from the electron ring that their influence on the equilibrium configuration can be ignored.
3. The minor dimensions of the ring are much smaller than its major radius, that is, the ring is *thin* with

$$a, b \ll R_0 \ . \tag{3.5.2}$$

4. It is further assumed that

$$\frac{\nu}{\gamma_0} = \frac{N_e}{2\pi R_0} \frac{e^2}{m_e c^2} \frac{1}{\gamma_0} \ll 1 , \tag{3.5.3}$$

where ν is Budker's parameter,[30] N_e is the total number of electrons in the ring, $-e$ is the electron charge, m_e is the electron rest mass, c is the speed of light *in vacuo*, $e^2/m_e c^2 = r_0$ is the classical electron radius, and $\gamma_0 m_e c^2$ is the characteristic energy of an electron in the ring.

Assumptions 3 and 4, which are consistent with parameters in existing electron ring accelerator experiments, ensure that the self fields are sufficiently weak that the equilibrium ring radius R_0 is equal, in a first approximation, to the Larmor radius of a single electron with energy $\gamma_0 m_e c^2$. It is informative to rewrite the inequality in Eq. (3.5.3) in a notation familiar in plasma physics. For a thin ring with elliptical cross section (minor dimensions equal to $2a$ and $2b$), the characteristic electron density is $\bar{n}_e = N_e/2\pi R_0 \pi ab$. It is straightforward to show that Eq. (3.5.3) can be expressed in the equivalent form

$$\frac{\nu}{\gamma_0} = \frac{ab}{4} \frac{\bar{\omega}_{pe}^2}{c^2 \gamma_0} \ll 1 , \tag{3.5.4}$$

where $\bar{\omega}_{pe}^2/\gamma_0 = 4\pi \bar{n}_e e^2/m_e \gamma_0$ is the electron plasma frequency-squared in the laboratory frame. Equation (3.5.4) and hence Eq. (3.5.3) are simply statements that the minor dimensions of the ring are small in comparison with the collisionless skin depth $c/\bar{\omega}_{pe}\gamma_0^{-1/2}$.

Central to a Vlasov-Maxwell description of relativistic ring equilibria are the single-particle constants of the motion in the equilibrium fields (see Section 3.1.3). For the equilibrium configuration illustrated in Fig. 3.5.1, the single-particle constants of the motion in the equilibrium fields are the total energy H and the canonical angular momentum P_θ, where H and P_θ can be expressed as

$$H = (m_e^2 c^4 + c^2 \mathbf{p}^2)^{1/2} - e\phi^0(r, z) , \tag{3.5.5}$$

$$P_\theta = r p_\theta - \frac{er}{c} A_\theta^{\text{ext}}(r, z) - \frac{er}{c} A_\theta^s(r, z) , \tag{3.5.6}$$

where $\phi^0(r, z)$ is the equilibrium electrostatic potential, $A_\theta^{\text{ext}}(r, z)$ is the θ-component of vector potential for the external confining field, $A_\theta^s(r, z)$ is the θ-component of vector potential for the equilibrium self magnetic field generated by the azimuthal electron current, $\mathbf{p}$ is the mechanical momentum, and $\mathbf{p}^2 \equiv p_r^2 + p_z^2 + p_\theta^2$. As discussed in Section 3.1.3, any distribution function $f_e^0(H, P_\theta)$ that is a function only of the single-particle constants of the motion, H and P_θ, is a solution to the steady-state ($\partial/\partial t = 0$) Vlasov equation for the electrons. Once the functional form of $f_e^0(H, P_\theta)$ is specified, the electrostatic potential $\phi^0(r, z)$ can be determined self-consistently from the equilibrium Poisson equation, Eq. (3.1.33). For a partially neutralizing ion background with density $n_i^0(r, z) = fn_e^0(r, z)$ [Eq. (3.5.1)], Eq. (3.1.33) reduces to

$$\frac{1}{r}\frac{\partial}{\partial r} r \frac{\partial}{\partial r}\phi^0(r, z) + \frac{\partial^2}{\partial z^2}\phi^0(r, z) = 4\pi e(1 - f)\int d^3p\, f_e^0(H, P_\theta)\,, \tag{3.5.7}$$

where $\int d^3p\, f_e^0(H, P_\theta)$ is related to the electron density $n_e^0(r, z)$ by

$$n_e^0(r, z) = \int d^3p\, f_e^0(H, P_\theta)\,, \tag{3.5.8}$$

and single ionization has been assumed in Eq. (3.5.7). Furthermore, the θ-component of the vector potential for the equilibrium self magnetic field is determined self-consistently from Eq. (3.1.34). Since it is assumed that the equilibrium ion current is equal to zero in the laboratory frame, Eq. (3.1.34) reduces to

$$\frac{\partial}{\partial r}\frac{1}{r}\frac{\partial}{\partial r} rA_\theta^s(r, z) + \frac{\partial^2}{\partial z^2}A_\theta^s(r, z) = \frac{4\pi e}{c}\int d^3p\, v_\theta f_e^0(H, P_\theta)\,, \tag{3.5.9}$$

where $v_\theta = (p_\theta/m_e)(1 + \mathbf{p}^2/m_e^2c^2)^{-1/2}$. In Eq. (3.5.9), $-e\int d^3p\, v_\theta f_e^0(H, P_\theta)$ is related to the azimuthal electron current $J_\theta^0(r, z)$ by

$$J_\theta^0(r, z) = -en_e^0(r, z)V_{e\theta}^0(r, z) = -e\int d^3p\, v_\theta f_e^0(H, P_\theta)\,, \tag{3.5.10}$$

where $V_{e\theta}^0(r, z)$ is the equilibrium azimuthal velocity of an electron fluid element.† In general, the determination of the equilibrium self-field potentials,

---

†Since $f_e^0(H, P_\theta)$ is an even function of p_r and p_z [see Eqs. (3.5.5) and (3.5.6)], it follows that

$$\int d^3p\, v_r f_e^0(H, P_\theta) = 0 = \int d^3p\, v_z f_e^0(H, P_\theta),$$

where $v_j \equiv (p_j/m_e)(1 + \mathbf{p}^2/m_e^2c^2)^{-1/2}$. Therefore the equilibrium flux of electrons in the r- and z-directions is equal to zero, as expected.

$\phi^0(r, z)$ and $A_\theta^s(r, z)$, from Eqs. (3.5.7) and (3.5.9) is not entirely straightforward. Since H and P_θ are functions of $\phi^0(r, z)$ and $A_\theta^s(r, z)$ [see Eqs. (3.5.5) and (3.5.6)], the charge density and current density in Eqs. (3.5.7) and (3.5.9) are *also* functions of $\phi^0(r, z)$ and $A_\theta^s(r, z)$. Except for certain simple choice of $f_e^0(H, P_\theta)$, Eqs. (3.5.7) and (3.5.9) are generally *nonlinear* equations for the self-field potentials. Fortunately, however, an approximate evaluation of the equilibrium properties of the ring is tractable for a thin ring with $\nu/\gamma_0 \ll 1$ [Eqs. (3.5.2) and (3.5.3)], which is the regime of experimental interest for electron ring accelerators.

3.5.2 Example of an Electron Ring Equilibrium with Thermal Energy Spread

There is considerable latitude in the choice of equilibrium distribution function $f_e^0(H, P_\theta)$. Therefore the choice of $f_e^0(H, P_\theta)$ should be guided by what appears appropriate in a given experimental situation. As an example that illustrates the essential features of the equilibrium analysis, consider the equilibrium distribution function specified by[132]

$$f_e^0(H, P_\theta) = N_0\, \delta(P_\theta - P_0) \exp\left[-(H - \gamma_0 m_e c^2 + e\bar{\phi}_0)/\Theta_e\right], \tag{3.5.11}$$

where $\bar{\phi}_0 = $ const., H and P_θ are defined in Eqs. (3.5.5) and (3.5.6), and N_0, P_0, γ_0, and Θ_e are positive constants. For the distribution function defined in Eq. (3.5.11), note that all of the electrons have the same value of canonical angular momentum ($P_\theta = P_0$). However, a thermal spread in energy H is incorporated in the factor $\exp[-(H - \gamma_0 m_e c^2 + e\bar{\phi}_0)/\Theta_e]$. Since the δ-function in Eq. (3.5.11) selects $P_\theta = P_0$, it follows from Eq. (3.5.6) that electrons located at (r, z) have azimuthal momentum

$$p_\theta = \frac{P_0}{r} + \frac{eA^0(r, z)}{c}, \tag{3.5.12}$$

where

$$A^0(r, z) \equiv A_\theta^{\text{ext}}(r, z) + A_\theta^s(r, z). \tag{3.5.13}$$

Without loss of generality, in Eq. (3.5.11) $\gamma_0 m_e c^2$ is taken to be the azimuthal energy of an electron located at $(r, z) = (R_0, 0)$ with zero transverse momentum ($p_\perp^2 = p_r^2 + p_z^2 = 0$), that is,

$$\gamma_0 m_e c^2 \equiv \left\{ m_e^2 c^4 + c^2 \left[\frac{P_0}{R_0} + \frac{eA^0(R_0, 0)}{c}\right]^2 \right\}^{1/2}.$$

Furthermore, $\bar{\phi}_0$ is taken to be the value of $\phi^0(r, z)$ at $(r, z) = (R_0, 0)$,

$$\bar{\phi}_0 \equiv \phi^0(R_0, 0). \tag{3.5.14}$$

Since the ring is thin, the transverse (r, z) excursions of the electrons composing the ring are small in comparison with R_0 [see Eq. (3.5.2)], and the exponent in Eq. (3.5.11) can be expanded with $c^2 p_\perp^2 \ll \gamma_\theta^2(r, z)\, m_e^2 c^4$, where $p_\perp^2 \equiv p_r^2 + p_z^2$ is the transverse momentum-squared, and $\gamma_\theta(r, z) m_e c^2$ is the azimuthal energy,

$$\gamma_\theta(r, z) m_e c^2 \equiv \left\{ m_e^2 c^4 + c^2 \left[\frac{P_0}{r} + \frac{eA^0(r, z)}{c} \right]^2 \right\}^{1/2}. \tag{3.5.15}$$

Approximating $H \simeq \gamma_\theta(r, z) m_e c^2 + (2m_e)^{-1} p_\perp^2/\gamma_\theta(r, z) - e\phi^0(r, z)$ [see Eq. (3.5.5)], we can express the equilibrium distribution function defined in Eq. (3.5.11) as

$$f_e^0(H, P_\theta) = N_0\, \delta(P_\theta - P_0) \exp\left\{ \frac{-p_\perp^2}{2\gamma_\theta(r, z) m_e \Theta_e} \right\} \exp\left\{ -\frac{\psi(r, z)}{\Theta_e} \right\}. \tag{3.5.16}$$

where

$$\psi(r, z) \equiv [\gamma_\theta(r, z) - \gamma_0] m_e c^2 - e\, \delta\phi^0(r, z) \tag{3.5.17}$$

and

$$\delta\phi^0(r, z) \equiv \phi^0(r, z) - \bar{\phi}_0 . \tag{3.5.18}$$

Equation (3.5.16) is a sufficiently accurate representation of $f_e^0(H, P_\theta)$ for present purposes. The approximate form of H used in obtaining Eq. (3.5.16) is valid provided the transverse kinetic energy, $p_\perp^2/2m_e\gamma_\theta(r, z)$ is small in comparison with the azimuthal energy, $\gamma_\theta(r, z) m_e c^2$. It is straightforward to show that this inequality is satisfied (in an average sense) provided $\Theta_e \ll \gamma_\theta(r, z) m_e c^2$, which is assumed to be the case.† For future reference, note from Eq. (3.5.17) that $\psi(R_0, 0) = 0$, since $\delta\phi^0(R_0, 0) = 0$ and $\gamma_\theta(R_0, 0) = \gamma_0$.

The macroscopic equilibrium properties of the electron ring, for example, density profile, azimuthal velocity profile, and transverse temperature profile,

† The condition for the average transverse kinetic energy to be small in comparison with the azimuthal energy is $\langle p_\perp^2 \rangle / 2m_e\gamma_\theta(r, z) \ll \gamma_\theta(r, z) m_e c^2$. Combining this inequality with Eqs. (3.5.21) and (3.5.23) gives $\Theta_e \ll \gamma_\theta(r, z) m_e c^2$.

can be determined by calculating the appropriate momentum moments of Eq. (3.5.16). For example, substituting Eq. (3.5.16) into Eq. (3.5.8) and representing $\int d^3p = 2\pi \int_{-\infty}^{\infty} dp_\theta \int_0^\infty dp_\perp p_\perp$, we can express the electron density $n_e^0(r, z)$ as

$$n_e^0(r,z) = \bar{n}_e \frac{R_0}{r} \frac{\gamma_\theta(r,z)}{\gamma_0} \exp\left\{-\frac{\psi(r,z)}{\Theta_e}\right\}. \tag{3.5.19}$$

where $\bar{n}_e \equiv n_e^0(R_0, 0)$ is the electron density at $(r, z) = (R_0, 0)$. [In terms of $\bar{n}_e$, the normalization constant in Eq. (3.5.16) is $N_0 = \bar{n}_e R_0/(2\pi\Theta_e\gamma_0 m_e)$.] The azimuthal velocity of an electron fluid element, $V_{e\theta}^0(r, z)$, and the transverse temperature profile, $T_{e\perp}^0(r, z)$, are defined in terms of the equilibrium distribution function, $f_e^0(H, P_\theta)$, by†

$$V_{e\theta}^0(r,z) \equiv \frac{\displaystyle\int d^3p \frac{c^2 p_\theta}{(m_e^2 c^4 + c^2 \mathbf{p}^2)^{1/2}} f_e^0(H, P_\theta)}{\displaystyle\int d^3p\, f_e^0(H, P_\theta)} \tag{3.5.20}$$

and

$$T_{e\perp}^0(r,z) \equiv \frac{\displaystyle\frac{1}{2}\int d^3p \frac{c^2 p_\perp^2}{(m_e^2 c^4 + c^2 \mathbf{p}^2)^{1/2}} f_e^0(H, P_\theta)}{\displaystyle\int d^3p\, f_e^0(H, P_\theta)}. \tag{3.5.21}$$

Substituting Eq. (3.5.16) into Eqs. (3.5.20) and (3.5.21) gives

$$V_{e\theta}^0(r,z) = \frac{1}{m_e\gamma_\theta(r,z)}\left[\frac{P_0}{r} + \frac{e}{c}A^0(r,z)\right] \tag{3.5.22}$$

---

†For the equilibrium distribution function defined in Eq. (3.5.11), it can be shown from Eq. (1.3.22) that the equilibrium stress tensor for the electrons is of the form

$$P_e^0(\mathbf{x}) = n_e^0(r,z) T_{e\perp}^0(r,z)\,(\hat{\mathbf{e}}_r\hat{\mathbf{e}}_r + \hat{\mathbf{e}}_z\hat{\mathbf{e}}_z),$$

where $n_e^0(r, z)$ and $T_{e\perp}^0(r, z)$ are defined in Eqs. (3.5.8) and (3.5.21), and $\hat{\mathbf{e}}_r$ and $\hat{\mathbf{e}}_z$ are unit vectors in the r- and z-directions, respectively. This result follows since $f_e^0(H, P_\theta)$ is an even function of p_r and p_z [see Eqs. (3.5.5) and (3.5.6)] and is symmetric under interchange of p_r^2 and p_z^2. Note that the electrons are *cold* in the θ-direction, since $[P_e^0(\mathbf{x})]_{\theta\theta} = 0$ for the choice of distribution function in Eq. (3.5.11).

and

$$T^0_{e\perp}(r, z) = \Theta_e. \tag{3.5.23}$$

In obtaining Eqs. (3.5.22) and (3.5.23), the approximation

$$\{m_e^2 c^4 + c^2 [P_0 / r + eA^0(r, z) / c]^2 + c^2 p_\perp^2\}^{-1/2} \simeq [\gamma_\theta(r, z) m_e c^2]^{-1}$$

has been made in the integrands of Eqs. (3.5.20) and (3.5.21) [see Eq. (3.5.15) and the discussion preceding it]. The errors incurred in Eqs. (3.5.22) and (3.5.23) by this approximation are of the order $\Theta_e/\gamma_\theta(r, z)mc^2 \ll 1$. For a thin ring with equilibrium distribution function given by Eq. (3.5.16), it follows from Eq. (3.5.23) that the transverse temperature profile is *isothermal* with temperature $T^0_{e\perp}(r, z) = \Theta_e =$ const. Note from Eqs. (3.5.19) and (3.5.22) that the equilibrium density and velocity profiles, $n_e^0(r, z)$ and $V^0_{e\theta}(r, z)$, are determined in terms of the self-field potentials, $\phi^0(r, z)$ and $A^s_\theta(r, z)$, and other properties characteristic of the equilibrium distribution function, for example, $P_0, \gamma_0 m_e c^2$, and $A_\theta^{\text{ext}}(r, z)$. The self-field potentials may be calculated self-consistently by substituting Eqs. (3.5.19) and (3.5.22) into Eqs. (3.5.7) and (3.5.9), and solving the resulting equations for $\phi^0(r, z)$ and $A^s_\theta(r, z)$.

3.5.3 Electron Density Profile

It is of considerable interest to determine a closed expression for the *r–z* dependence of the electron density profile $n_e^0(r, z)$. For a thin ring with $\nu/\gamma_0 \ll 1$, the variation of $\gamma_\theta(r, z)$ and $1/r$ across the minor dimensions of the ring is small. Therefore Eq. (3.5.19) can be expressed in the approximate form

$$n_e^0(r, z) = \bar{n}_e \exp\left\{-\frac{\psi(r, z)}{\Theta_e}\right\}. \tag{3.5.24}$$

The detailed shape of the density profile is contained in the factor $\exp\{-\psi(r, z)/\Theta_e\}$ in Eq. (3.5.24). Following Davidson, Lawson, and Mahajan,[25,132] we Taylor expand the expression for $\psi(r, z)$ given in Eq. (3.5.17) about $(r, z) = (R_0, 0)$ for $a, b \ll R_0$ and $\nu/\gamma_0 \ll 1$. Introducing $\rho = r - R_0$, and neglecting terms higher than quadratic order, we can express

$$\begin{aligned}\psi(r, z) = {} & \rho\left[\frac{\partial}{\partial r}\psi(r, z)\right]_{R_0,0} + z\left[\frac{\partial}{\partial z}\psi(r, z)\right]_{R_0,0} + \rho z\left[\frac{\partial^2}{\partial r\,\partial z}\psi(r, z)\right]_{R_0,0} \\ & + \frac{\rho^2}{2}\left[\frac{\partial^2}{\partial r^2}\psi(r, z)\right]_{R_0,0} + \frac{z^2}{2}\left[\frac{\partial^2}{\partial z^2}\psi(r, z)\right]_{R_0,0} + \dots,\end{aligned} \tag{3.5.25}$$

where use has been made of $\psi(R_0, 0) = 0$ [see the discussion following Eq. (3.5.18)].

The evaluation of the coefficients in Eq. (3.5.25) closely parallels the analysis in Section IV and Appendix A of Reference 25. Therefore, for present purposes, it is adequate to outline the results, and the reader is referred to Reference 25 for further details on procedure. First, it can be shown from Eq. (3.5.17) that $[\partial\psi(r, z)/\partial z]_{R_0,0} = 0$ and $[\partial^2 \psi(r, z)/\partial r\ \partial z]_{R_0,0} = 0$ follow directly from axial symmetry of the equilibrium configuration about $z = 0$. [In particular, the coefficients of z and ρz vanish in Eq. (3.5.25) since the radial magnetic field and axial electric field are identically zero in the midplane, that is, $B_r^0(r, 0) = 0 = E_z^0(r, 0)$, where $B_r^0(r, z) = -\partial A^0(r, z)/\partial z$ and $E_z^0(r, z) = -\partial\ \delta\phi^0(r, z)/\partial z$.] Second, the requirement that $[\partial\psi(r, z)/\partial r]_{R_0,0} = 0$ is imposed, which assures that the term linear in ρ is absent in Eq. (3.5.25). This condition effectively *determines* the equilibrium radius R_0 that corresponds to the geometric center of the beam. Making use of Eq. (3.5.17) and the definitions of radial electric field $[E_r^0(r, z) = -\partial\ \delta\phi^0(r, z)/\partial r]$ and axial magnetic field $[B_z^0(r, z) = \partial A^0(r, z)/\partial r + A^0(r, z)/r]$, we can express the condition $[\partial\psi(r, z)/\partial r]_{R_0,0} = 0$ in the equivalent form

$$-\frac{\gamma_0 m_e \beta_\theta^2 c^2}{R_0} = -eE_r^0(R_0, 0) - e\beta_\theta B_z^0(R_0, 0)\,, \tag{3.5.26}$$

where $\beta_\theta c \equiv V_{e\theta}^0(R_0, 0) = m_e^{-1}\gamma_0^{-1}\ [P_0/R_0 + eA^0(R_0, 0)/c]$ is the azimuthal velocity of an electron fluid element located at $(r, z) = (R_0, 0)$ [see Eq. (3.5.22)]. Equation (3.5.26), which is a statement of radial force balance on an electron fluid element at $(r, z) = (R_0, 0)$, effectively determines the equilibrium radius R_0 of the ring. For a thin ring with $\nu/\gamma_0 \ll 1$, the self-field contributions in Eq. (3.5.26) are small in comparison with $B_z^{\text{ext}}(R_0, 0)$, and Eq. (3.5.26) can be approximated by[144]

$$-\frac{\gamma_0 m_e \beta_\theta^2 c^2}{R_0} = -e\beta_\theta B_z^{\text{ext}}(R_0, 0) \tag{3.5.27}$$

correct to lowest order. Since the terms in Eq. (3.5.25) that are proportional to ρ, z, and ρz vanish, Eq. (3.5.25) reduces to

$$\psi(r, z) = \frac{\rho^2}{2}\left[\frac{\partial^2}{\partial r^2}\psi(r, z)\right]_{R_0,0} + \frac{z^2}{2}\left[\frac{\partial^2}{\partial z^2}\psi(r, z)\right]_{R_0,0}, \tag{3.5.28}$$

where terms higher than quadratic order have been neglected. When Eq. (3.5.28) is substituted into Eq. (3.5.24), it is found that $n_e^0(r, z)$ has a Gaussian profile about $(r, z) = (R_0, 0)$, that is,

$$n_e^0(r, z) = \bar{n}_e \exp\left\{-\frac{\rho^2}{2a^2} - \frac{z^2}{2b^2}\right\}, \tag{3.5.29}$$

where $\rho = r - R_0$, and

$$\frac{1}{a^2} \equiv \frac{1}{\Theta_e}\left[\frac{\partial^2}{\partial r^2}\psi(r, z)\right]_{R_0,0}, \tag{3.5.30}$$

$$\frac{1}{b^2} \equiv \frac{1}{\Theta_e}\left[\frac{\partial^2}{\partial z^2}\psi(r, z)\right]_{R_0,0}. \tag{3.5.31}$$

To evaluate a^2 and b^2, the analysis in Reference 25 is paralleled. For a thin ring with $\nu/\gamma_0 \ll 1$, it is straightforward to show from Eqs. (3.5.17), (3.5.30), and (3.5.31) that

$$\frac{1}{a^2} = \frac{\gamma_0 m_e c^2}{\Theta_e}\frac{\beta_\theta^2}{R_0^2}\left\{1 - n + \frac{R_0^2 e}{\gamma_0 \beta_\theta^2 m_e c^2}\left[\frac{\partial}{\partial r}E_r^0(r, z) + \beta_\theta \frac{\partial}{\partial r}B_z^s(r, z)\right]_{R_0,0}\right\} \tag{3.5.32}$$

and

$$\frac{1}{b^2} = \frac{\gamma_0 m_e c^2}{\Theta_e}\frac{\beta_\theta^2}{R_0^2}\left\{n + \frac{R_0^2 e}{\gamma_0 \beta_\theta^2 m_e c^2}\left[\frac{\partial}{\partial z}E_z^0(r, z) - \beta_\theta \frac{\partial}{\partial z}B_r^s(r, z)\right]_{R_0,0}\right\}, \tag{3.5.33}$$

where n is the external field index at $(r, z) = (R_0, 0)$,

$$n \equiv -\left[\frac{r}{B_z^{\text{ext}}(r, z)}\frac{\partial}{\partial r}B_z^{\text{ext}}(r, z)\right]_{R_0,0}. \tag{3.5.34}$$

Since the self fields are weak for $\nu/\gamma_0 \ll 1$, terms in Eqs. (3.5.30) and (3.5.31) that are proportional to $E_r^0(R_0, 0)$ and $B_z^s(R_0, 0)$ have been consistently neglected in obtaining Eqs. (3.5.32) and (3.5.33). Terms that are proportional to *gradients* of the self fields, however, are retained in the analysis, for example, $[\partial B_z^s(r, z)/\partial r]_{R_0,0}$ and $[\partial E_r^0(r, z)/\partial r]_{R_0,0}$.

In order to obtain *closed* expressions for a and b it is necessary to evaluate the self-field gradient terms in Eqs. (3.5.32) and (3.5.33), a procedure that requires a self-consistent determination of $\phi^0(r, z)$ and $A_\theta^s(r, z)$ from Eqs. (3.5.7) and (3.5.9). Considerable simplification occurs in the analysis since $a, b \ll R_0$, $\nu/\gamma_0 \ll 1$, and only the self-field gradients *evaluated at* $(R_0, 0)$ are required in Eqs. (3.5.32) and (3.5.33). In Poisson's equation [Eq. (3.5.7)], use is made of the expression for $n_e^0(r, z)$ given in Eq. (3.5.29). Moreover, since $V_{e\theta}^0(r, z)$ varies only a small amount across the minor dimensions of the ring, the approximation $V_{e\theta}^0(r, z) \simeq \beta_\theta c \equiv V_\theta^0(R_0, 0) = \gamma_0^{-1} m_e^{-1} [P_0/R_0 + eA^0(R_0, 0)/c]$ [see Eq. (3.5.22)] is made in the $\nabla \times \mathbf{B}_0^s$ Maxwell equation [Eq. (3.5.9)]. Introducing the variable $\rho = r - R_0$, we can then express Eqs. (3.5.7) and (3.5.9) as

$$\left(\frac{\partial^2}{\partial \rho^2} + \frac{\partial^2}{\partial z^2}\right) \phi^0(\rho, z) = 4\pi e(1 - f)\bar{n}_e \exp\left(-\frac{\rho^2}{2a^2} - \frac{z^2}{2b^2}\right), \quad (3.5.35)$$

and

$$\left(\frac{\partial^2}{\partial \rho^2} + \frac{\partial^2}{\partial z^2}\right) A_\theta^s(\rho, z) = 4\pi e\beta_\theta \bar{n}_e \exp\left(-\frac{\rho^2}{2a^2} - \frac{z^2}{2b^2}\right), \quad (3.5.36)$$

where the differential operators on the left-hand sides of Eqs. (3.5.7) and (3.5.9) have been approximated by their limiting values for $b/R_0, a/R_0 \to 0$. In terms of a Fourier integral representation, the solution to Eq. (3.5.35) is

$$\phi^0(\rho, z) = -4\pi e(1 - f)\frac{ab}{2\pi}\bar{n}_e \int_{-\infty}^{\infty} dk \int_{-\infty}^{\infty} d\ell \exp(ikz + i\ell\rho)$$
$$\times \frac{\exp(-a^2\ell^2/2 - b^2k^2/2)}{\ell^2 + k^2}, \quad (3.5.37)$$

and the solution to Eq. (3.5.36) can be expressed in a similar form with $(1 - f)$ replaced by β_θ. Making use of the integral representations of $\phi^0(\rho, z)$ and $A_\theta^s(\rho, z)$, it is straightforward to evaluate the self-field gradient terms in Eqs. (3.5.32) and (3.5.33). For example, it follows from Eq. (3.5.37) that

$$\left[\frac{\partial}{\partial z}E_z^0(r, z)\right]_{R_0,0} = -\left[\frac{\partial^2}{\partial z^2}\phi^0(\rho, z)\right]_{0,0}$$
$$= -4\pi e(1 - f)\frac{ab}{2\pi}\bar{n}_e \int_{-\infty}^{\infty} dk \int_{-\infty}^{\infty} d\ell \frac{k^2 \exp(-a^2\ell^2/2 - b^2k^2/2)}{\ell^2 + k^2}.$$
$$(3.5.38)$$

Carrying out the integrations in Eq. (3.5.38) gives

$$\left[\frac{\partial}{\partial z}E_z^0(r,z)\right]_{R_0,0} = -4\pi e(1-f)\bar{n}_e \frac{a}{a+b}. \tag{3.5.39}$$

In a similar fashion it can be shown that

$$\left[\frac{\partial}{\partial r}E_r^0(r,z)\right]_{R_0,0} = -4\pi e(1-f)\bar{n}_e \frac{b}{a+b}, \tag{3.5.40}$$

$$\left[\frac{\partial}{\partial z}B_r^s(r,z)\right]_{R_0,0} = -4\pi e\beta_\theta \bar{n}_e \frac{a}{a+b}, \tag{3.5.41}$$

$$\left[\frac{\partial}{\partial r}B_z^s(r,z)\right]_{R_0,0} = 4\pi e\beta_\theta \bar{n}_e \frac{b}{a+b}, \tag{3.5.42}$$

correct to lowest order in a/R_0, b/R_0. Substituting Eqs. (3.5.39)-(3.5.42) into Eqs. (3.5.32) and (3.5.33) gives[132]

$$\frac{1}{a^2} = \frac{\gamma_0 m_e c^2}{\Theta_e}\frac{\beta_\theta^2}{R_0^2}\left[1-n+2\frac{\nu}{\gamma_0}\frac{R_0^2}{a(a+b)\beta_\theta^2}(\beta_\theta^2+f-1)\right], \tag{3.5.43}$$

and

$$\frac{1}{b^2} = \frac{\gamma_0 m_e c^2}{\Theta_e}\frac{\beta_\theta^2}{R_0^2}\left[n+2\frac{\nu}{\gamma_0}\frac{R_0^2}{b(a+b)\beta_\theta^2}(\beta_\theta^2+f-1)\right], \tag{3.5.44}$$

where $\nu \equiv (N_e/2\pi R_0)(e^2/m_e c^2)$ is Budker's parameter, and $N_e = (2\pi R_0) \times 2(\pi ab)\bar{n}_e = 2\pi R_0 \bar{n}_e \int\int d\rho\, dz \exp(-\rho^2/2a^2 - z^2/2b^2)$ is the total number of electrons in the ring. Equations (3.5.43) and (3.5.44) constitute closed equations for a and b in terms of properties of the equilibrium distribution function (e.g., $\Theta_e, \gamma_0, \beta_\theta$) and the external field configuration (e.g., n).

The expressions for a and b given in Eqs. (3.5.43) and (3.5.44) correspond to the radial and axial betatron oscillation amplitudes[145, 146] (including equilibrium self-field effects) for an electron with transverse energy equal to the *thermal* energy Θ_e. Note that the thin ring approximation [Eq. (3.5.3)] is valid only if the energy spread of the electrons is small, that is, $\Theta_e \ll \gamma_0 m_e c^2$. Furthermore, the self-field contributions in Eqs. (3.5.43) and (3.5.44) [the terms proportional to (β_θ^2+f-1)] are in the direction of focusing the beam, that is, *decreasing* a^2 and b^2, only if

$$\beta_\theta^2 > 1-f. \tag{3.5.45}$$

Equation (3.5.45) is the familiar condition that magnetic pinching forces exceed electrostatic repulsive forces.[95, 96]

The sensitive dependence of equilibrium properties on the choice of $f_e^0(H, P_\theta)$ should be noted. In Reference 25, where all the electrons are assumed to have the same value of total energy H and the same value of canonical angular momentum P_θ, that is, $f_e^0(H, P_\theta) = N_0\,\delta(H - \gamma_0 mc^2 + e\bar{\phi}_0)\,\delta(P_\theta - P_0)$, it is found that the transverse temperature profile is parabolic with $T_{e\perp}^0(r, z)$ assuming its maximum value at $(r, z) = (R_0, 0)$. Furthermore, in Reference 25 the electron density is approximately constant in the ring interior, and the minor cross section of the ring has a sharp boundary with envelope equation $\rho^2/a^2 + z^2/b^2 = 1$.† This is in contrast to the results obtained for the equilibrium distribution function in Eq. (3.5.11), where the transverse electron temperature profile is found to be isothermal [Eq. (3.5.23)], and the minor cross section of the ring has a *diffuse* boundary [Eq. (3.5.29)]. It may be anticipated that the stability properties of these two equilibrium distribution functions are also quite different.

In conclusion, it is important to note that the assumption that the ion density profile satisfies $n_i^0(r, z) = fn_e^0(r, z)$ is highly idealized and probably is not satisfied in many applications of interest. In this regard it is straightforward to extend the present equilibrium analysis to describe the positive ion background within the framework of the steady-state Vlasov-Maxwell equations. If the ions are described by an equilibrium distribution function $f_i^0(H)$,‡ the procedure for calculating the equilibrium properties of the electron ring from $f_e^0(H, P_\theta)$ remains essentially the same. The only difference is that the electrostatic potential $\phi^0(r, z)$ must be calculated self-consistently, using the ion density computed from $f_i^0(H)$. In other words, the right-hand side of Eq. (3.5.7) must be replaced by

$$-4\pi\rho^0(r, z) = 4\pi e\left[\int d^3p\, f_e^0(H, P_\theta) - n_i^0(r, z)\right] \tag{3.5.46}$$

where $n_i^0(r, z) \equiv \int d^3p\, f_i^0(H)$.

---

†Here a and b are defined in Eqs. (66) and (67) of Reference 25.

---

‡See Section VI of Reference 25. Note from Eq. (3.1.30) that H is an even function of p_r, p_θ, and p_z. Therefore the mean ion velocity is

$$\mathbf{V}_i^0(\mathbf{x}) = \int d^3p\, \mathbf{v} f_i^0(H) = 0$$

(see assumption 1). If the ion dynamics are nonrelativistic, then

$$H = \mathbf{p}^2 / 2m_i + e\phi^0(r, z) + m_i c^2$$

follows from Eq. (3.1.30), where m_i and $+e$ are ion mass and charge, respectively.

3.6 STABILITY THEOREM FOR NONRELATIVISTIC NONDIAMAGNETIC EQUILIBRIA

The detailed stability properties of spatially nonuniform Vlasov equilibria with equilibrium self fields are usually difficult to ascertain analytically. Therefore even a sufficient condition for stability is a welcome result since it provides valuable information regarding the class of equilibrium distribution functions that may be unstable. In this section a sufficient condition for stability[77] is derived for the nonrelativistic, nondiamagnetic Vlasov equilibria discussed in Section 3.2. The equilibrium configuration consists of a nonneutral plasma column aligned parallel to a uniform external magnetic field, $\mathbf{B}_0^{\text{ext}}(\mathbf{x}) = B_0\hat{\mathbf{e}}_z$ (see Fig. 3.1.1). As in Section 3.2, the equilibrium properties are assumed to be independent of z ($\partial/\partial z = 0$) and azimuthally symmetric ($\partial/\partial\theta = 0$) about an axis of symmetry parallel to $B_0\hat{\mathbf{e}}_z$, and the influence of external boundaries is ignored. Furthermore, it is assumed that the particle motions are nonrelativistic, and that the axial and azimuthal self magnetic fields, $B_z^s(r)$ and $B_\theta^s(r)$, are negligibly small. For simplicity, it is also assumed that the nonneutral plasma column is composed only of electrons.† In general, the equilibrium distribution function for the electrons is of the form

$$f_e^0(\mathbf{x}, \mathbf{p}) = f_e^0(H, P_\theta, p_z)\,, \tag{3.6.1}$$

where $p_z = m_e v_z$ is the axial momentum, and

$$H = \frac{\mathbf{p}^2}{2m_e} - e\phi^0(r)\,, \tag{3.6.2}$$

$$P_\theta = r\left(p_\theta - m_e r\frac{\Omega_e}{2}\right). \tag{3.6.3}$$

The notation used in Eqs. (3.6.1)-(3.6.3) is the same as in Section 3.2, and the equilibrium electrostatic potential $\phi^0(r)$ is determined self-consistently in terms of $f_e^0(H, P_\theta, p_z)$ from the equilibrium Poisson equation, Eq. (3.2.5).

To determine the stability properties of the equilibrium distribution function, $f_e^0(H, P_\theta, p_z)$, the time development of perturbations about equilibrium is examined within the framework of the Vlasov-Maxwell equations (see Section 1.3.2). For the equilibrium configuration considered here, the electron distribution function, $f_e(\mathbf{x}, \mathbf{p}, t)$, electric field, $\mathbf{E}(\mathbf{x}, t)$, and magnetic field, $\mathbf{B}(\mathbf{x}, t)$, can be expressed as

$$f_e(\mathbf{x}, \mathbf{p}, t) = f_e^0(H, P_\theta, p_z) + \delta f_e(\mathbf{x}, \mathbf{p}, t)\,, \tag{3.6.4}$$

---

†The analysis can be extended in a straightforward manner to include ions (see the discussion at the end of Section 3.6).

$$\mathbf{E}(\mathbf{x}, t) = E_r^0(r)\hat{\mathbf{e}}_r + \delta\mathbf{E}(\mathbf{x}, t) , \tag{3.6.5}$$

$$\mathbf{B}(\mathbf{x}, t) = B_0 \hat{\mathbf{e}}_z + \delta\mathbf{B}(\mathbf{x}, t) , \tag{3.6.6}$$

where $\hat{\mathbf{e}}_r$ and $\hat{\mathbf{e}}_z$ are unit vectors in the r- and z-directions (see Fig. 3.1.1), $E_r^0(r) = -\partial\phi^0(r)/\partial r$ is the equilibrium radial electric field, and the equilibrium self magnetic fields have been neglected in Eq. (3.6.6). In the present analysis, a sufficient condition for stability of the equilibrium distribution function $f_e^0(H, P_\theta, p_z)$ is derived in the *electrostatic approximation*. In other words, it is assumed that the perturbed magnetic field, $\delta\mathbf{B}(\mathbf{x}, t)$, remains negligibly small as the system evolves, and the $\nabla \times \mathbf{E}$ Maxwell equation, Eq. (1.3.4), is approximated by

$$\nabla \times \mathbf{E}(\mathbf{x}, t) = 0 . \tag{3.6.7}$$

The analysis can be extended in a relatively straightforward manner to include the perturbed magnetic field $\delta\mathbf{B}(\mathbf{x}, t)$. (See the discussion at the end of Section 3.6.) Approximating $\mathbf{B}(\mathbf{x}, t) \simeq B_0\hat{\mathbf{e}}_z$ in Eq. (1.3.2), we can express the Vlasov equation for $f_e(\mathbf{x}, \mathbf{p}, t)$ as

$$\left\{ \frac{\partial}{\partial t} + \mathbf{v} \cdot \frac{\partial}{\partial \mathbf{x}} - e\left[\mathbf{E}(\mathbf{x}, t) + \frac{\mathbf{v} \times B_0 \hat{\mathbf{e}}_z}{c}\right] \cdot \frac{\partial}{\partial \mathbf{p}} \right\} f_e(\mathbf{x}, \mathbf{p}, t) = 0 , \tag{3.6.8}$$

where $\mathbf{v} = \mathbf{p}/m_e$, since the particle motions are nonrelativistic. The electric field $\mathbf{E}(\mathbf{x}, t)$ in Eq. (3.6.8) is determined self-consistently in terms of $f_e(\mathbf{x}, \mathbf{p}, t)$ from Poisson's equation, Eq. (1.3.6). Since no ions are present, Eq. (1.3.6) reduces to

$$\nabla \cdot \mathbf{E}(\mathbf{x}, t) = -4\pi e \int d^3p\, f_e(\mathbf{x}, \mathbf{p}, t) , \tag{3.6.9}$$

where $-e$ is the electron charge, and $\rho_{\text{ext}}(\mathbf{x}) = 0$ is assumed. Note that Eq. (3.6.8) is fully nonlinear, that is, no small-amplitude approximation has been made in this equation.

To derive a sufficient condition for stability, consider the function $F(t)$, defined by[77]

$$F(t) = \int d^3x \left\{ \frac{\mathbf{E}^2 - \mathbf{E}^{02}}{8\pi} + \int d^3p \times \left[\left(\frac{\mathbf{p}^2}{2m_e} - \omega_e P_\theta \right) \left(f_e - f_e^0 \right) + G(f_e) - G(f_e^0) \right] \right\} , \tag{3.6.10}$$

where spatial integrations are over the infinite domain,† and G is a smooth and differentiable (but otherwise arbitrary) function of its argument f_e. In Eq. (3.6.10), ω_e = const., P_θ is defined in Eq. (3.6.3), and the abbreviated notation, $\mathbf{E} = \mathbf{E}(\mathbf{x}, t)$, $\mathbf{E}^0 = E_r^0(r)\hat{\mathbf{e}}_r$, $f_e = f_e(\mathbf{x}, \mathbf{p}, t)$, and $f_e^0 = f_e^0(H, P_\theta, p_z)$, has been introduced. Making use of Eqs. (3.6.7)-(3.6.10) and some straightforward integration by parts, we can show that

$$\frac{d}{dt}F(t) = -\omega_e \int \frac{d^3x}{4\pi}\, \hat{\mathbf{e}}_z \cdot \mathbf{x} \times \mathbf{E}\, \nabla \cdot \mathbf{E}$$

$$= -\omega_e \int \frac{d^3x}{4\pi}\, \hat{\mathbf{e}}_z \cdot \nabla \cdot \left[\mathbf{x} \times \left(\mathbf{E}\mathbf{E} - \frac{\mathbf{E}^2}{2} I \right) \right]^T \tag{3.6.11}$$

where $[\;]^T$ denotes diadic transpose. The right-hand side of Eq. (3.6.11) integrates to zero provided the perturbed electric field, $\delta\mathbf{E} = \mathbf{E} - \mathbf{E}^0$, vanishes sufficiently rapidly as $\mathbf{x} \to \infty$. Therefore $dF(t)/dt = 0$, that is,

$$F(t) = F(0) = \text{const.} \tag{3.6.12}$$

Note that the constancy of F is an exact consequence of the fully nonlinear Vlasov-Poisson equations.

Now consider small-amplitude perturbations. Taylor expanding $G(f_e) = G(f_e^0 + \delta f_e)$ for small δf_e gives

$$G(f_e) = G(f_e^0) + G'(f_e^0)\,\delta f_e + G''(f_e^0)\frac{(\delta f_e)^2}{2} + \cdots . \tag{3.6.13}$$

Correct to second order in the perturbation amplitude, Eq. (3.6.10) can be expressed as

$$F^{(2)} = \int d^3x \left(\frac{(\delta\mathbf{E})^2}{8\pi} + \int d^3p \times \left\{ \left[\frac{\mathbf{p}^2}{2m_e} - e\phi^0(r) - \omega_e P_\theta + G'(f_e^0) \right] \delta f_e + G''(f_e^0)\frac{(\delta f_e)^2}{2} \right\} \right), \tag{3.6.14}$$

---

†If $f_e(\mathbf{x}, \mathbf{p}, t)$ and $\mathbf{E}(\mathbf{x}, t)$ are spatially periodic in the z-direction with periodicity length $2L$, then

$$\int d^3x = \int_{-\infty}^{\infty} dx \int_{-\infty}^{\infty} dy \int_{-L}^{L} dz$$

in Eq. (3.6.10).

where $\delta\mathbf{E} = \mathbf{E} - \mathbf{E}^0$ and $\delta f_e = f_e - f_e^0$. In obtaining Eq. (3.6.14) from Eq. (3.6.10), use has been made of Eq. (3.6.13) and the identity

$$\int \frac{d^3x}{4\pi} \mathbf{E}^0 \cdot \delta\mathbf{E} = \int \frac{d^3x}{4\pi} \phi^0 \nabla \cdot \delta\mathbf{E} = -e \int d^3x\, \phi^0 \int d^3p\, \delta f_e\,, \quad (3.6.15)$$

which follows from $\mathbf{E}^0 = -\nabla\phi^0$ and $\nabla \cdot \delta\mathbf{E} = -4\pi e \int d^3p\, \delta f_e$. The function $G(f_e^0)$, which has been arbitrary up to this point, is now chosen to satisfy

$$G'(f_e^0) = -(H - \omega_e P_\theta)\,, \quad (3.6.16)$$

where $H = \mathbf{p}^2/2m_e - e\phi^0(r)$. The choice of $G'(f_e^0)$ in Eq. (3.6.16) implies that f_e^0 depends on H and P_θ through the linear combination $H - \omega_e P_\theta$, that is, the analysis is restricted to rigid-rotor equilibria with $f_e^0(H, P_\theta, p_z) = f_e^0(H - \omega_e P_\theta)$. Making use of Eq. (3.6.16), we find that the term linear in δf_e vanishes in Eq. (3.6.14), and $F^{(2)}$ reduces to

$$F^{(2)} = \int d^3x \left[\frac{(\delta\mathbf{E})^2}{8\pi} + \int d^3p\, G''(f_e^0) \frac{(\delta f_e)^2}{2} \right]. \quad (3.6.17)$$

Differentiating Eq. (3.6.16) with respect to f_e^0 gives

$$G''(f_e^0) = -\frac{1}{\partial f_e^0/\partial(H - \omega_e P_\theta)}. \quad (3.6.18)$$

Substituting Eq. (3.6.18) into Eq. (3.6.17), we can express $F^{(2)}$ in the equivalent form

$$F^{(2)} = \int d^3x \left\{ \frac{(\delta\mathbf{E})^2}{8\pi} + \int d^3p \frac{(\delta f_e)^2}{2} \left[\frac{-1}{\partial f_e^0/\partial(H - \omega_e P_\theta)} \right] \right\}. \quad (3.6.19)$$

If

$$\frac{\partial}{\partial(H - \omega_e P_\theta)} f_e^0(H - \omega_e P_\theta) \leqslant 0\,, \quad (3.6.20)$$

it follows from Eq. (3.6.19) that $F^{(2)}$ is a sum of nonnegative terms. Since F is a constant, the perturbations $\delta\mathbf{E}(\mathbf{x}, t)$ and $\delta f_e(\mathbf{x}, \mathbf{p}, t)$ cannot grow without bound when Eq. (3.6.20) is satisfied. Therefore, a sufficient condition for stability can be stated as follows:

> If $f_e^0(H - \omega_e P_\theta)$ is a monotonically decreasing function of $H - \omega_e P_\theta$, the equilibrium is stable to small-amplitude electrostatic perturbations. (3.6.21)

Equation (3.6.21) is the generalization of Newcomb's theorem [147] to a *non-neutral, rotating* plasma. The sufficient condition for stability stated in Eq. (3.6.21) is especially significant since it is applicable to spatially nonuniform equilibria characterized by an equilibrium self electric field, $\mathbf{E}^0(\mathbf{x}) = -\hat{\mathbf{e}}_r\, \partial\phi^0(r)/\partial r$. The stability theorem is applicable to surface perturbations as well as perturbations interior to the electron gas column. As an example, note that the Gibbs distribution function [Eq. (3.2.32)],

$$f_e^0(H-\omega_e P_\theta) = \frac{\bar{n}_e}{(2\pi m_e \Theta_e)^{3/2}} \exp\left\{-(H-\omega_e P_\theta)\,/\,\Theta_e\right\}\,, \quad (3.6.22)$$

is electrostatically stable within the context of the present analysis. The loss-cone distribution function specified in Eq. (3.2.21), however, *may* be unstable since it is *not* a monotonically decreasing function of $H - \omega_e P_\theta$ (see Section 3.7.3).

Several important generalizations of the preceding analysis can be made. First, if the analysis is extended to include a perturbed magnetic field $\delta\mathbf{B}(\mathbf{x}, t)$ the same stability condition is obtained.[148] In other words, $\partial f_e^0(H-\omega_e P_\theta)/\partial(H-\omega_e P_\theta) \leqslant 0$ is a sufficient condition for stability of $f_e^0(H-\omega_e P_\theta)$ to small-amplitude electromagnetic perturbations with arbitrary polarization. Second, the stability theorem is readily extended to a multicomponent nonneutral plasma provided each component is rotating with the same angular velocity, $\omega_\alpha = \omega_0 = \text{const}$. It is found that† $\partial f_\alpha^0(H-\omega_0 P_\theta)/\partial(H-\omega_0 P_\theta) \leqslant 0$, for each α, is a sufficient condition for stability of $f_\alpha^0(H-\omega_0 P_\theta)$ to small-amplitude perturbations. Finally, following Gardner,[149] the stability theorem can be extended to show that $\partial f_\alpha^0(H-\omega_0 P_\theta)/\partial(H-\omega_0 P_\theta) \leqslant 0$, for each α, is a sufficient condition for *nonlinear* stability of $f_\alpha^0(H-\omega_0 P_\theta)$ to arbitrary-amplitude perturbations.

3.7 BODY WAVES IN A NONNEUTRAL PLASMA COLUMN

3.7.1 Equilibrium Configuration and Assumptions

In this section the time-dependent Vlasov-Maxwell equations[76, 77] are used to study the dispersive properties of small-amplitude perturbations propagating in

---

†Keep in mind that

$$H = \mathbf{p}^2\,/\,2m_\alpha + e_\alpha\phi^0(r) \qquad \text{and} \qquad P_\theta = r\,(p_\theta + m_\alpha r \epsilon_\alpha \Omega_\alpha/\,2)$$

for a particle of species α.

the interior of a constant-density nonneutral plasma column aligned parallel to a uniform external magnetic field $\mathbf{B}_0^{\text{ext}}(\mathbf{x}) = B_0 \hat{\mathbf{e}}_z$ (Fig. 3.1.1). The notation and general assumptions regarding the equilibrium configuration (i.e., azimuthal symmetry, nonrelativistic particle dynamics, negligible diamagnetism, etc.) are the same as in Section 3.2 (see also Section 3.6). It is further assumed that perturbations are about *rigid-rotor* equilibria of the form discussed in Section 3.2.2, that is, the equilibrium distribution function for the electrons is of the form

$$f_e^0(\mathbf{x}, \mathbf{p}) = f_e^0(H - \omega_e P_\theta, p_z) , \tag{3.7.1}$$

where

$$H = \frac{\mathbf{p}^2}{2m_e} - e\phi^0(r), \tag{3.7.2}$$

$$P_\theta = r \left(p_\theta - m_e r \frac{\Omega_e}{2}\right) , \tag{3.7.3}$$

$p_z = m_e v_z$ is the axial momentum, and ω_e = const. = angular velocity of mean rotation. For simplicity, the stability analysis is carried out for a pure electron plasma. The resulting dispersion relation is then generalized to a multicomponent nonneutral plasma.

As discussed in Section 3.2.2, if ω_e is closely tuned to the laminar rotation velocity ω_e^+ or ω_e^- [Eq. (3.2.19)], the electron density profile $n_e^0(r)$ has a characteristic radial dimension R_p much larger than a thermal electron Debye length, and the electron density is approximately constant in the column interior (Fig. 3.7.1). In other words, if

$$\omega_e = \omega_e^+(1 - \delta) \qquad \text{or} \qquad \omega_e = \omega_e^-(1 + \delta), \quad 0 < \delta \ll 1, \tag{3.7.4}$$

where

$$\omega_e^\pm = \frac{\Omega_e}{2}\left[1 \pm \left(1 - \frac{2\overline{\omega}_{pe}^2}{\Omega_e^2}\right)^{1/2}\right], \tag{3.7.5}$$

then the electron density profile can be approximated by

$$n_e^0(r) = \begin{cases} \overline{n}_e = \text{const.}, & 0 < r \lesssim R_p , \\ 0, & r \gtrsim R_p , \end{cases} \tag{3.7.6}$$

for equilibrium distribution functions of the form $f_e^0(H - \omega_e P_\theta, p_z)$. In Eq. (3.7.5), $\Omega_e = eB_0/m_e c$ and $\overline{\omega}_{pe}^2 = 4\pi \overline{n}_e e^2/m_e$. For constant-density rigid-rotor equilibria with $\omega_e \simeq \omega_e^+$ or $\omega_e \simeq \omega_e^-$ [Eq. (3.7.4)], it is relatively straightforward to determine the detailed stability properties of $f_e^0(H - \omega_e P_\theta, p_z)$, assuming small-amplitude perturbations localized to the column interior ($r < R_p$), with characteristic transverse wavelength $|k_\perp|^{-1}$ small in comparison with the column radius R_p,

$$k_\perp^2 R_p^2 \gg 1 . \quad (3.7.7)$$

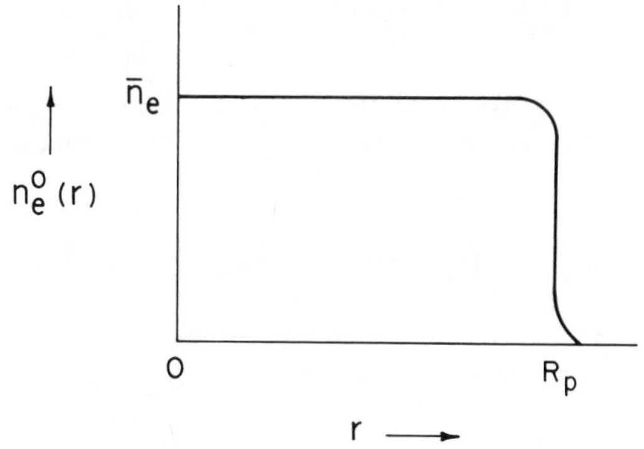

Fig. 3.7.1 Plot of $n_e^0(r)$ versus r for nonneutral rigid-rotor equilibria of the form $f_e^0(H - \omega_e P_\theta, p_z)$ for the case in which ω_e is closely tuned to the laminar rotation velocity ω_e^+ or ω_e^- [Eq. (3.7.4)]. The electron density is approximately constant in the column interior.

For such perturbations the analysis of the linearized Vlasov-Maxwell equations simplifies considerably since the column radius is effectively infinite, and boundary conditions at $r \simeq R_p$ can be ignored.

Wave perturbations inside the plasma are known as *body waves*. Before deriving the dispersion relation for body waves in a nonneutral plasma column, it is useful to summarize some additional properties of the equilibrium. In general, the linear combination $H - \omega_e P_\theta$ that occurs in Eq. (3.7.1) can be expressed as

$$H - \omega_e P_\theta = [p_r^2 + (p_\theta - m_e r\omega_e)^2 + p_z^2] / 2m_e$$

$$+ \frac{m_e}{2}\left[r^2(\omega_e \Omega_e - \omega_e^2) - \frac{2e}{m_e}\phi^0(r)\right] , \quad (3.7.8)$$

where use has been made of Eqs. (3.7.2) and (3.7.3). For a constant-density column [Eq. (3.7.6)], the equilibrium Poisson equation can be integrated to give

$$\phi^0(r) = \frac{m_e \bar{\omega}_{pe}^2}{4e} r^2 \tag{3.7.9}$$

for $0 < r < R_p$. Moreover, the equilibrium electric field $\mathbf{E}^0(\mathbf{x})$ in the region $0 < r < R_p$ can be expressed as

$$\mathbf{E}^0(\mathbf{x}) = -\nabla\phi^0(r) = -\frac{m_e \bar{\omega}_{pe}^2}{2e} r \,\hat{\mathbf{e}}_r \tag{3.7.10}$$

where $\hat{\mathbf{e}}_r$ is a unit vector in the r-direction (see Fig. 3.1.1). Substituting Eq. (3.7.9) into Eq. (3.7.8), and making use of Eq. (3.7.5), we can express $H - \omega_e P_\theta$ as

$$\begin{aligned} H - \omega_e P_\theta &= [p_r^2 + (p_\theta - m_e r \omega_e)^2 + p_z^2] / 2m_e \\ &\quad - \frac{m_e}{2}(\omega_e - \omega_e^+)(\omega_e - \omega_e^-) r^2 \end{aligned} \tag{3.7.11}$$

for $0 < r < R_p$. Since $\omega_e \simeq \omega_e^+$ or $\omega_e \simeq \omega_e^-$ for a constant-density column, Eq. (3.7.11) reduces to

$$H - \omega_e P_\theta = [p_r^2 + (p_\theta - m_e r \omega_e)^2 + p_z^2] / 2m_e \tag{3.7.12}$$

in the column interior.

In summary, for the subclass of rigid-rotor Vlasov equilibria with $\omega_e \simeq \omega_e^+$ or $\omega_e \simeq \omega_e^-$ [Eq. (3.7.4)], the electron density is constant in the column interior [Eq. (3.7.6)], the radial electric field varies linearly with radial distance r from the axis of rotation [Eq. (3.7.10)], and $H - \omega_e P_\theta$ can be identified with the kinetic energy of an electron in a frame of reference rotating with angular velocity $\omega_e = \text{const.}$ [Eq. (3.7.12)].

3.7.2 Electrostatic Dispersion Relation

To determine the stability properties of the equilibrium distribution function $f_e^0(H - \omega_e P_\theta, p_z)$, the time development of perturbations about equilibrium is examined within the framework of the Vlasov-Maxwell equations (see Sections 1.3.2 and 3.6). In this section the dispersion relation for body waves in a constant-density nonneutral plasma column is derived in the electrostatic approximation.[76, 77] In other words, it is assumed that the perturbed magnetic field

$\delta\mathbf{B}(\mathbf{x},t) = \mathbf{B}(\mathbf{x},t) - B_0\hat{\mathbf{e}}_z$ remains negligibly small as the system evolves, and the $\nabla\times\delta\mathbf{E}$ Maxwell equation, Eq. (1.3.15), can be approximated by

$$\nabla\times\delta\mathbf{E}(\mathbf{x},t) = 0\,, \tag{3.7.13}$$

where $\delta\mathbf{E}(\mathbf{x},t) = \mathbf{E}(\mathbf{x},t) - \mathbf{E}^0(\mathbf{x})$ is the perturbed electric field.† Approximating $\mathbf{B}(\mathbf{x},t) \simeq B_0\hat{\mathbf{e}}_z$ in Eq. (1.3.14), we can express the *linearized* Vlasov equation for the perturbed distribution function $\delta f_e(\mathbf{x},\mathbf{p},t) = f_e(\mathbf{x},\mathbf{p},t) - f_e^0\,(H-\omega_e P_\theta, p_z)$ as

$$\left\{\frac{\partial}{\partial t} + \mathbf{v}\cdot\frac{\partial}{\partial\mathbf{x}} - e\left[\mathbf{E}^0(\mathbf{x}) + \frac{\mathbf{v}\times B_0\hat{\mathbf{e}}_z}{c}\right]\cdot\frac{\partial}{\partial\mathbf{p}}\right\}\delta f_e(\mathbf{x},\mathbf{p},t)$$

$$= e\,\delta\mathbf{E}(\mathbf{x},t)\cdot\frac{\partial}{\partial\mathbf{p}} f_e^0(H-\omega_e P_\theta, p_z)\,, \tag{3.7.14}$$

where $\mathbf{v} = \mathbf{p}/m_e$ (since the particle motions are nonrelativistic). The perturbed electric field $\delta\mathbf{E}(\mathbf{x},t)$ in Eq. (3.7.14) is determined self-consistently in terms of $\delta f_e(\mathbf{x},\mathbf{p},t)$ from Poisson's equation, Eq. (1.3.17). Since no ions are present, Eq. (1.3.17) reduces to

$$\Delta\cdot\delta\mathbf{E}(\mathbf{x},t) = -4\pi e\int d^3p\;\delta f_e(\mathbf{x},\mathbf{p},t). \tag{3.7.15}$$

Note from Eq. (3.7.13) that $\delta\mathbf{E}(\mathbf{x},t)$ can be expressed as

$$\delta\mathbf{E}(\mathbf{x},t) = -\nabla\,\delta\phi(\mathbf{x},t) \tag{3.7.16}$$

in Eqs. (3.7.14) and (3.7.15). In obtaining Eq. (3.7.14), it has *not* been assumed that the electron density is constant in the column interior or that $\omega_e \simeq \omega_e^\pm$. Equations (3.7.14)-(3.7.16) are applicable for small-amplitude electrostatic perturbations about arbitrary rigid-rotor equilibria characterized by the equilibrium distribution function $f_e^0(H-\omega_e P_\theta, p_z)$ and the self-consistent electric field $\mathbf{E}^0(\mathbf{x})$.‡

∇

---

†Keep in mind that the equilibrium electric field $\mathbf{E}^0(\mathbf{x})$ is curl-free, $\nabla\times\mathbf{E}^0(\mathbf{x}) = 0$ [Eq. (1.3.9)].

---

‡In general, $\mathbf{E}^0(\mathbf{x}) = -\hat{\mathbf{e}}_r\,\partial\phi^0(r)/\partial r$ in Eq. (3.7.14), where $\phi^0(r)$ is determined self-consistently from

$$\frac{1}{r}\frac{\partial}{\partial r}r\frac{\partial}{\partial r}\phi^0(r) = 4\pi e\int d^3p\, f_e^0(H-\omega_e P_\theta, p_z)\,.$$

In circumstances where the electron density is constant in the column interior, Davidson and Krall have integrated Eq. (3.7.14), using the method of characteristics.[77] This approach involves a detailed calculation of the particle trajectories in the combined equilibrium electric field $\mathbf{E}^0(\mathbf{x})$ [Eq. (3.7.10)] and magnetic field $B_0\hat{\mathbf{e}}_z$. As an alternative approach, whereby the electrostatic dispersion relation can be obtained directly by analogy with the neutral plasma case, it is useful to transform Eqs. (3.7.14) and (3.7.15) to a frame of reference rotating with angular velocity $\omega_e = \text{const.}$ about the axis of symmetry (Fig. 3.7.2). Introducing cylindrical polar coordinates, we transform Eqs. (3.7.14) and (3.7.15) from the independent variables $(\mathbf{x}, \mathbf{p}, t)$ in the laboratory frame to the independent variables $(\mathbf{x}', \mathbf{p}', t')$ in the rotating frame, where

$$\begin{gathered} r' = r, \qquad \theta' = \theta - \omega_e t, \qquad z' = z, \\ p'_r = p_r, \qquad p'_\theta = p_\theta - m_e r \omega_e, \qquad p'_z = p_z, \\ t' = t. \end{gathered} \tag{3.7.17}$$

Making use of Eqs. (3.7.14)-(3.7.17), it is straightforward to show that the linearized Vlasov-Poisson equations in variables appropriate to the rotating frame can be expressed as[9]

$$\begin{aligned} &\left\{\frac{\partial}{\partial t'} + \mathbf{v}' \cdot \frac{\partial}{\partial \mathbf{x}'} - e\left[\mathbf{E}^0(\mathbf{x}') - \frac{m_e}{e}(\omega_e^2 - \omega_e \Omega_e)\mathbf{x}'_\perp\right] \cdot \frac{\partial}{\partial \mathbf{p}'}\right\} \delta f_e(\mathbf{x}', \mathbf{p}', t') \\ &\qquad - m_e(\Omega_e - 2\omega_e)\, \mathbf{v}' \times \hat{\mathbf{e}}'_z \cdot \frac{\partial}{\partial \mathbf{p}'} \delta f_e(\mathbf{x}', \mathbf{p}', t) \\ &\qquad = -e \nabla' \delta\phi(\mathbf{x}', t') \cdot \frac{\partial}{\partial \mathbf{p}'} f_e^0(H' - \omega_e P'_\theta, p'_z), \end{aligned} \tag{3.7.18}$$

and

$$\nabla'^2 \delta\phi(\mathbf{x}', t') = 4\pi e \int d^3p'\, \delta f_e(\mathbf{x}', \mathbf{p}', t'), \tag{3.7.19}$$

where $\mathbf{v}' = \mathbf{p}'/m_e$. In Eqs. (3.7.18) and (3.7.19), $\nabla' = \partial/\partial \mathbf{x}'$, and $\mathbf{x}'_\perp = r'\hat{\mathbf{e}}'_r$, where $\hat{\mathbf{e}}'_r$ is a unit vector in the r'-direction (in the rotating frame). Note from Eq. (3.7.18) that the effective equilibrium electric field in the rotating frame is

$$\mathbf{E}^0_{\text{eff}}(\mathbf{x}') = \mathbf{E}^0(\mathbf{x}') - \frac{m_e}{e}(\omega_e^2 - \omega_e \Omega_e)\mathbf{x}'_\perp. \tag{3.7.20}$$

The contribution, $m_e 2\omega_e \mathbf{v}' \times \hat{\mathbf{e}}'_z \cdot (\partial/\partial \mathbf{p}')\,\delta f_e$, in Eq. (3.7.18) arises from the Coriolis acceleration of the electrons.

Equation (3.7.18) simplifies considerably inside the plasma column $(0<r<R_p)$ when $\omega_e \simeq \omega_e^{\pm}$ [Eq. (3.7.4)] and the electron density is constant [Eq. (3.7.6)]. In this case, $\mathbf{E}^0(\mathbf{x}') = -(m_e\overline{\omega}_{pe}^2/2e)\mathbf{x}'_\perp$ [Eq. (3.7.10)], and $\mathbf{E}^0_{\text{eff}}(\mathbf{x}')$ reduces to

$$\mathbf{E}^0_{\text{eff}}(\mathbf{x}') = -\frac{m_e}{e}(\omega_e - \omega_e^+)(\omega_e - \omega_e^-)\mathbf{x}'_\perp = 0\,, \tag{3.7.21}$$

Furthermore, comparing Eqs. (3.7.12) and (3.7.17), we find that $H' - \omega_e P'_\theta$ is equal to the kinetic energy of an electron in the rotating frame, that is,

$$H' - \omega_e P'_\theta = \mathbf{p}'^2 / 2m_e\,, \tag{3.7.22}$$

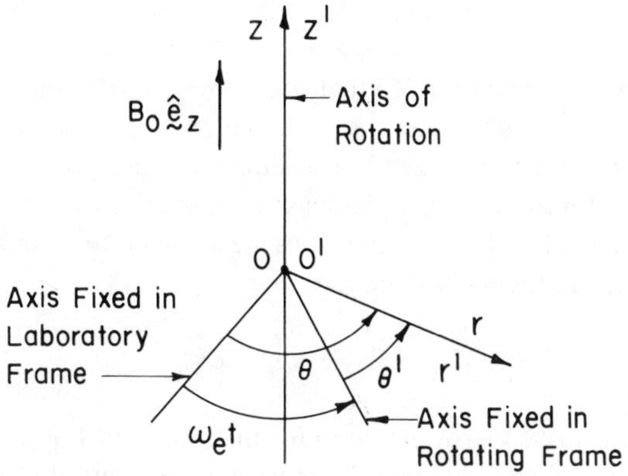

Fig. 3.7.2 Equations (3.7.14) and (3.7.15) are transformed to a frame of reference rotating with angular velocity ω_e = const. about the axis of symmetry. The variables (r', θ', z') in the rotating frame are related to the variables (r, θ, z) in the laboratory frame by $r' = r$, $\theta' = \theta - \omega_e t$, and $z' = z$.

where $\mathbf{p}'^2 = p_r'^2 + p_\theta'^2 + p_z'^2$. The advantage of transforming to the rotating frame when $\omega_e \simeq \omega_e^{\pm}$ is evident. Inside the plasma column the effective equilibrium electric field is equal to zero [Eq. (3.7.21)], and the equilibrium distribution function in the rotating frame is spatially uniform since

$$f_e^0(H' - \omega_e P'_\theta, p'_z) = f_e^0(\mathbf{p}'^2, p'_z) \tag{3.7.23}$$

is independent of r' [Eq. (3.7.22)].† When Eqs. (3.7.21) and (3.7.23) are substituted into Eq. (3.7.18), the linearized Vlasov equation in the rotating frame reduces to

$$\left\{\frac{\partial}{\partial t'} + \mathbf{v}' \cdot \frac{\partial}{\partial \mathbf{x}'} - m_e(\Omega_e - 2\omega_e)\mathbf{v}' \times \hat{\mathbf{e}}_z' \cdot \frac{\partial}{\partial \mathbf{p}'}\right\} \delta f_e(\mathbf{x}', \mathbf{p}', t')$$

$$= -e\nabla' \, \delta\phi(\mathbf{x}', t') \cdot \frac{\partial}{\partial \mathbf{p}'} f_e^0(\mathbf{p}'^2, p_z') \qquad (3.7.24)$$

for $0 < r < R_p$, and $\omega_e \simeq \omega_e^\pm$. Equation (3.7.24) is indeed a plausible result. Note that the only equilibrium force that the electrons experience in the rotating frame is equal to $-m_e(\Omega_e - 2\omega_e)\mathbf{v}' \times \hat{\mathbf{e}}_z'$. As discussed in Section 1.2, the perturbed orbits in the equilibrium fields are circular gyrations with angular gyration frequency equal to the vortex frequency,[8]

$$\omega_{ev} = \Omega_e - 2\omega_e \,. \qquad (3.7.25)$$

Note from Eqs. (3.7.5) and (3.7.25) that $\omega_{ev} = (\omega_e^+ - \omega_e^-)$ when $\omega_e = \omega_e^-$, and $\omega_{ev} = -(\omega_e^+ - \omega_e^-)$ when $\omega_e = \omega_e^+$. Equations (3.7.19) and (3.7.24) are similar in form to the linearized Vlasov-Poisson equations for electrostatic perturbations in a uniform neutral plasma immersed in an external magnetic field $B_0 \hat{\mathbf{e}}_z$. In particular, if the positive ions are assumed to form a fixed background ($m_i \rightarrow \infty$), and the replacement

$$\Omega_e \rightarrow \Omega_e - 2\omega_e = \omega_{ev} \qquad (3.7.26)$$

is made in the linearized Vlasov equation for the electrons, Eqs. (3.7.19) and (3.7.24) are recovered directly from the corresponding neutral plasma equations.[77]

It is straightforward to obtain the dispersion relation for electrostatic perturbations in a constant-density electron plasma by Fourier-Laplace transforming Eqs. (3.7.19) and (3.7.24) with respect to x' and t'. Alternatively, the appropriate dispersion relation can be written down by analogy with the neutral plasma result; it is this approach that is used in the present analysis. First, assume that the perturbations $\delta f_e(\mathbf{x}', \mathbf{p}', t)$ and $\delta\phi(\mathbf{x}, t')$ in Eqs. (3.7.19) and (3.7.24) are proportional to

$$\exp\,[i(\mathbf{k}' \cdot \mathbf{x}' - \omega' t')] \,. \qquad (3.7.27)$$

---

†Only the dependence of $H' - \omega_e P_\theta'$ on $\mathbf{p}'^2$ has been displayed on the right-hand side of Eq. (3.7.23), and the factor $(2m_e)^{-1}$ has been omitted [see Eq. (3.7.22)].

Then make the replacement $\Omega_e \to \Omega_e - 2\omega_e = \omega_{ev}$ [Eq. (3.7.26)] in the neutral plasma dispersion relation,[150] assuming a fixed ion background ($m_i \to \infty$). The resulting dispersion relation for a pure electron plasma is

$$0 = D(k'_\perp, k'_z, \omega') = 1 + \frac{4\pi e^2}{m_e k'^2} \sum_{n=-\infty}^{\infty} \int_L d^3p'\, J_n^2\left(\frac{k'_\perp v'_\perp}{\omega_e^+ - \omega_e^-}\right)$$

$$\times \frac{\left[\dfrac{n(\omega_e^+ - \omega_e^-)}{v'_\perp}\dfrac{\partial}{\partial v'_\perp} + k'_z \dfrac{\partial}{\partial v'_z}\right] f_e^0(\mathbf{p}'^2, p'_z)}{\omega' - k'_z v'_z - n(\omega_e^+ - \omega_e^-)}, \tag{3.7.28}$$

where L denotes the Landau contour.† In Eq. (3.7.28), $\mathbf{v}' = \mathbf{p}'/m_e$, $v'_z = \mathbf{v}' \cdot \hat{\mathbf{e}}'_z$, $v'_\perp = |\mathbf{v}'_\perp|$ (where $\mathbf{v}'_\perp = \mathbf{v}' - v'_z \hat{\mathbf{e}}'_z$), $k'_z = \mathbf{k}' \cdot \hat{\mathbf{e}}'_z$, $k'_\perp = |\mathbf{k}'_\perp|$ (where $\mathbf{k}'_\perp = \mathbf{k}' - k'_z \hat{\mathbf{e}}'_z$), $k'^2 = k'^2_\perp + k'^2_z$, and J_n is the Bessel function of the first kind of order n. In obtaining Eq. (3.7.28) use has been made of the identity $\omega_{ev} = \mp(\omega_e^\pm - \omega_e^-)$ for $\omega_e = \omega_e^\pm$ [see Eqs. (3.7.5) and (3.7.25)].‡

Equation (3.7.28), which relates the wave vector $\mathbf{k}'$ and the complex oscillation frequency ω', is the dispersion relation in a frame of reference rotating with angular velocity $\omega_e (\simeq \omega_e^\pm)$ about the axis of symmetry (see Fig. 3.7.1), that is, in a frame of reference corotating with the equilibrium. To obtain the dispersion relation appropriate to the laboratory frame, it is necessary to relate $(k'_\perp, k'_z, \omega')$ in the rotating frame to $(k_\perp, k_z, \omega)$ in the laboratory frame. For spatial perturbations with azimuthal harmonic number ℓ,‡ the relation is

---

†The Landau contour in Eq. (3.7.28) refers to the integration with respect to p'_z. If $\text{Im}\,\omega' > 0$, then $\int_L dp'_z = \int_{-\infty}^{\infty} dp'_z$, that is, the integration is along the real p'_z-axis. However, if $\text{Im}\,\omega' \leqslant 0$, then $\int_L dp'_z$ loops under the singularity in the integrand in the manner discussed in Sections 8.4 and 8.10 of Reference 88.

---

‡Note that the integrand in Eq. (3.7.28) remains unchanged for $(\omega_e^+ - \omega_e^-) \to -(\omega_e^+ - \omega_e^-)$ and $n \to -n$.

---

‡This refers to perturbations where the θ-dependence of $\delta f_e(\mathbf{x}, \mathbf{p}, t)$ and $\delta\phi(\mathbf{x}, t)$ is proportional to $\exp(i\ell\theta)$ in the laboratory frame.

$$\omega' = \omega - \ell\omega_e, \qquad k'_\perp = k_\perp, \qquad k'_z = k_z . \tag{3.7.29}$$

Since the equilibrium is azimuthally symmetric, the only effect is to Doppler-shift the frequency by $\ell\omega_e$. Dropping the prime notation on $\mathbf{p}'$ and $\mathbf{v}'$ in Eq. (3.7.28), and making use of Eq. (3.7.29), we can express the dispersion relation for a pure electron plasma in the laboratory frame as [76, 77]

$$0 = D_\ell(k_\perp, k_z, \omega) = 1 + \frac{4\pi e^2}{m_e k^2} \sum_{n=-\infty}^{\infty} \int_L d^3p \, J_n^2\left(\frac{k_\perp v_\perp}{\omega_e^+ - \omega_e^-}\right) \times \frac{\left[\dfrac{n(\omega_e^+ - \omega_e^-)}{v_\perp}\dfrac{\partial}{\partial v_\perp} + k_z \dfrac{\partial}{\partial v_z}\right] f_e^0(\mathbf{p}^2, p_z)}{\omega - \ell\omega_e - k_z v_z - n(\omega_e^+ - \omega_e^-)} . \tag{3.7.30}$$

Equation (3.7.30) relates $\mathbf{k}$ and ω for spatial perturbations with azimuthal harmonic number ℓ. Keep in mind the range of applicability of Eq. (3.7.30). First, it has been assumed that $\omega_e \simeq \omega_e^\pm$ [Eq. (3.7.4)] and that the electron density is constant in the column interior [Eq. (3.7.6)]. Second, boundary effects at $r \simeq R_p$ have been ignored, and the analysis is restricted to electrostatic perturbations localized to the column interior with $r < R_p$ and $k_\perp^2 R_p^2 \gg 1$. Finally, Eq. (3.7.30) is the dispersion relation for a pure electron plasma. It has been assumed that no ions are present in the system $[f_i^0(\mathbf{x}, \mathbf{p}) = 0]$, and that the plasma consists of a single component of electrons (rotating with mean angular velocity $\omega_e \simeq \omega_e^+$ or $\omega_e \simeq \omega_e^-$).

It is straightforward to extend the previous analysis to a multicomponent nonneutral plasma, assuming that the equilibrium distribution function for each plasma component is of the form $f_\alpha^0(H - \omega_\alpha P_\theta, p_z)$, and that ω_α is closely tuned to the laminar rotation velocity ω_α^+ or ω_α^-, where [see Eq. (2.2.12)]

$$\omega_\alpha^\pm = -\frac{\epsilon_\alpha \Omega_\alpha}{2}\left[1 \pm \left(1 - 2\,\frac{\sum_\eta 4\pi e_\alpha e_\eta \bar{n}_\eta / m_\alpha}{\Omega_\alpha^2}\right)^{1/2}\right]. \tag{3.7.31}$$

In Eq. (3.7.31), $\epsilon_\alpha = \text{sgn}\, e_\alpha$ and $\Omega_\alpha = |e_\alpha| B_0 / m_\alpha c$. If $\omega_\alpha \simeq \omega_\alpha^+$ or $\omega_\alpha \simeq \omega_\alpha^-$ for each plasma component, the density of each component is approximately constant $[n_\alpha^0(r) = \bar{n}_\alpha = \text{const.}]$ in the column interior, and $H' - \omega_\alpha P'_\theta = \mathbf{p}'^2/2m_\alpha$ in a frame of reference rotating with angular velocity ω_α. In the multicomponent case, the dispersion relation for electrostatic perturbations in the column interior can be expressed as

$$0 = D_\ell(k_\perp, k_z, \omega) = 1 + \sum_\alpha \frac{4\pi e_\alpha^2}{m_\alpha k^2} \sum_{n=-\infty}^{\infty} \int_L d^3p\, J_n^2\left(\frac{k_\perp v_\perp}{\omega_\alpha^+ - \omega_\alpha^-}\right)$$

$$\times \frac{\left[\dfrac{n(\omega_\alpha^+ - \omega_\alpha^-)}{v_\perp}\dfrac{\partial}{\partial v_\perp} + k_z \dfrac{\partial}{\partial v_z}\right] f_\alpha^0(\mathbf{p}^2, p_z)}{\omega - \ell\omega_\alpha - k_z v_z - n(\omega_\alpha^+ - \omega_\alpha^-)}. \qquad (3.7.32)$$

Equation (3.7.32) is the appropriate generalization of Eq. (3.7.30) to a multicomponent nonneutral plasma. It relates the wave vector **k** and complex oscillation frequency ω (in the laboratory frame) for spatial perturbations with azimuthal harmonic ℓ. If no ions are present in the system $[f_i^0(\mathbf{x}, \mathbf{p}) = 0]$, and the plasma consists of a single component of electrons (rotating with mean angular velocity $\omega_e = \omega_e^+$ or $\omega_e = \omega_e^-$), Eq. (3.7.32) reduces to Eq. (3.7.30). On the other hand, if the plasma is electrically neutral $(\Sigma_\eta e_\eta \bar{n}_\eta = 0)$, and each component is in the slow rotational mode $(\omega_\alpha = \omega_\alpha^-)$, Eq. (3.7.32) reduces to the familar dispersion relation for electrostatic perturbations in a uniform neutral plasma.[150] This follows since $\omega_\alpha^+ - \omega_\alpha^- = -\epsilon_\alpha \Omega_\alpha$, and $\omega_\alpha^- = 0$, for $\Sigma_\eta e_\eta \bar{n}_\eta = 0$ [see Eq. (3.7.31)].

Since Eqs. (3.7.30) and (3.7.32) are similar in structure to the corresponding dispersion relations for a neutral plasma, many of the waves and instabilities that depend on the detailed momentum-space structure of $f_\alpha^0(\mathbf{p}^2, p_z)$ have their analogs in a constant-density nonneutral plasma in circumstances where the present analysis is applicable. In fact, the results of a large body of neutral plasma literature can be applied virtually intact with the replacements $\omega \to \omega - \ell\omega_\alpha$ and $\Omega_\alpha \to \pm(\omega_\alpha^+ - \omega_\alpha^-)$. Specific examples are discussed in Section 3.7.3.

3.7.3 Examples of Electrostatic Waves and Instabilities

In this section some of the electrostatic waves and instabilities characteristic of body wave perturbations in a pure electron plasma are discussed.[76, 77] Use is made of the dispersion relation in Eq. (3.7.30) and of the algorithms for obtaining stability information for a pure electron plasma from the corresponding results for a neutral plasma, that is,

$$\omega \to \omega - \ell\omega_e, \qquad \Omega_e \to \pm(\omega_e^+ - \omega_e^-), \qquad m_i \to \infty. \qquad (3.7.33)$$

A. Electron Plasma Oscillations at Brillouin Flow

Equation (3.7.30) is valid for electron density in the range $0 < 2\bar{\omega}_{pe}^2/\Omega_e^2 < 1$. In the limit of Brillouin flow,[133] $2\bar{\omega}_{pe}^2/\Omega_e^2 = 1$, note from Eq. (3.7.5) and Fig. 3.2.1 that $\omega_e^\pm = \Omega_e/2$, and

$$\omega_e^+ - \omega_e^- = \Omega_e \left(1 - \frac{2\overline{\omega}_{pe}^2}{\Omega_e^2}\right) = 0 \,. \qquad (3.7.34)$$

From Eq. (3.7.33), the analogous limit in the neutral plasma case is the zero magnetic field limit, $\Omega_e \to 0$. Therefore, at Brillouin flow, the dispersion relation in Eq. (3.7.30) can be expressed as

$$0 = 1 - \frac{4\pi e^2}{m_e k^2} \int_L d^3p \, \frac{\mathbf{k} \cdot (\partial/\partial \mathbf{v})}{\omega' - \mathbf{k} \cdot \mathbf{v}} f_e^0(\mathbf{p}^2, p_z) \,, \qquad (3.7.35)$$

where $\omega' = \omega - \ell\Omega_e/2$, $\mathbf{k} = \mathbf{k}_\perp + k_z \hat{\mathbf{e}}_z$, and $\mathbf{v} = \mathbf{p}/m_e$. Except for the Doppler shift in frequency by $\ell\Omega_e/2$, Eq. (3.7.35) is identical to the dispersion relation for electrostatic perturbations in a uniform, *unmagnetized* neutral plasma,[151] ignoring the ion dynamics ($m_i \to \infty$). Depending on the detailed form of $f_e^0(\mathbf{p}^2, p_z)$, Eq. (3.7.35) can support solutions corresponding to instability† ($\mathrm{Im}\,\omega > 0$). These include the gentle bump-in-tail instability[152] and the non-resonant two-stream instability [e.g., if the p_z-dependence of $f_e^0(\mathbf{p}^2, p_z)$ corresponds to two cold, counterstreaming electron components.]

As an example corresponding to stable oscillations ($\mathrm{Im}\,\omega < 0$), consider the thermal equilibrium distribution function specified by [see Eq. (3.2.32)]

$$f_e^0(H - \omega_e P_\theta, p_z) = \frac{\overline{n}_e}{(2\pi m_e \Theta_e)^{3/2}} \exp\left[-(H - \omega_e P_\theta)/\Theta_e\right] . \qquad (3.7.36)$$

Comparing Eqs. (3.7.22), (3.7.23), and (3.7.36) for $\omega_e \simeq \omega_e^\pm$ and $0 < r < R_p$, we can express $f_e^0(\mathbf{p}^2, p_z)$ as

$$f_e^0(\mathbf{p}^2, p_z) = \frac{\overline{n}_e}{(2\pi m_e \Theta_e)^{3/2}} \exp(-\mathbf{p}^2/2m_e\Theta_e) \,. \qquad (3.7.37)$$

Dividing ω into its real and imaginary parts, $\omega = \omega_r + i\omega_i$, and substituting Eq. (3.7.37) into Eq. (3.7.35), we can show that

$$(\omega_r - \ell\Omega_e/2)^2 = \overline{\omega}_{pe}^2(1 + 3k^2\lambda_D + \cdots) \,, \qquad (3.7.38)$$

---

†It has been assumed that the time variation of perturbed quantities is proportional to $\exp(-i\omega t)$. Therefore $\mathrm{Im}\,\omega > 0$ corresponds to temporal growth.

and

$$\omega_i \simeq -\left(\frac{\pi}{8}\right)^{1/2} \frac{\bar{\omega}_{pe}}{(|\mathbf{k}|\lambda_D)^3} \exp\left(-\frac{1}{2k^2\lambda_D{}^2} - \frac{3}{2}\right) \tag{3.7.39}$$

for $k^2\lambda_D^2 = k^2(\Theta_e/4\pi\bar{n}_e e^2) \ll 1$. In other words, for wavelengths long in comparison with the electron Debye length λ_D, the system supports weakly damped $[|\omega_i/(\omega_r - \ell\Omega_e/2)| \ll 1]$ electron plasma oscillations.

B. Bernstein Modes

As a further example, consider Eq. (3.7.30) for $k_z = 0$, $\omega_e^+ - \omega_e^- = \Omega_e \times (1 - 2\bar{\omega}_{pe}^2/\Omega_e^2)^{1/2} \neq 0$, and the equilibrium distribution function specified by Eq. (3.7.37). In this case, Eq. (3.7.30) reduces to

$$0 = 1 - \frac{\bar{\omega}_{pe}^2}{(\omega_e^+ - \omega_e^-)^2} \sum_{n=-\infty}^{\infty} \frac{n^2(\omega_e^+ - \omega_e^-)^2}{(\omega - \ell\omega_e)^2 - n^2(\omega_e^+ - \omega_e^-)^2} \frac{\exp(-\lambda_e) I_n(\lambda_e)}{\lambda_e}, \tag{3.7.40}$$

where $\lambda_e = k_\perp^2\Theta_e/m_e(\omega_e^+ - \omega_e^-)^2$, and I_n is the modified Bessel function of the first kind of order n. Equation (3.7.40) is the analog of the Bernstein-mode dispersion relation[153] for a pure electron plasma. The solutions to Eq. (3.7.40) correspond to pure oscillations (Im $\omega = 0$). If the electron density is low, $\bar{\omega}_{pe}^2/(\omega_e^+ - \omega_e^-)^2 \ll 1$, the solutions to Eq. (3.7.40) can be approximated by

$$(\omega - \ell\omega_e)^2 = n^2(\omega_e^+ - \omega_e^-)^2[1 + \alpha_n(\lambda_e)], \quad n = \pm 1, \pm 2, \pm 3, \ldots, \tag{3.7.41}$$

where

$$\alpha_n(\lambda_e) = \frac{2\bar{\omega}_{pe}^2}{(\omega_e^+ - \omega_e^-)^2} \frac{\exp(-\lambda_e) I_n(\lambda_e)}{\lambda_e}, \tag{3.7.42}$$

and $\alpha_n(\lambda_e) \ll 1$. Therefore a low-density pure electron plasma, characterized by the equilibrium distribution function in Eq. (3.7.37), supports oscillations near harmonics of $(\omega_e^+ - \omega_e^-)$ for electrostatic waves propagating perpendicular to $B_0\hat{\mathbf{e}}_z$.

C. Loss-Cone Instability

The neutral plasma analog of Eq. (3.7.30) has been extensively investigated for equilibrium distribution functions of the loss-cone form.[154, 155] It may be anticipated that loss-cone distributions are also relevant to nonneutral plasma experiments carried out in mirror geometries. As an example, consider the equilibrium distribution function specified by [see Eq. (3.2.21)]

$$f_e^0(H-\omega_e P_\theta, p_z) = \frac{\overline{n}_e}{2\pi m_e}\,\delta(p_z)\,\delta(H-\omega_e P_\theta - m_e V_{e\perp}^{02}/2)\,. \quad (3.7.43)$$

Comparing Eqs. (3.7.22), (3.7.23), and (3.7.43) for $\omega_e \simeq \omega_e^{\pm}$ and $0<r<R_p$, we can express $f_e^0(\mathbf{p}^2, p_z)$ as

$$f_e^0(\mathbf{p}^2, p_z) = \frac{\overline{n}_e}{\pi}\,\delta(p_z)\,\delta\,[p_\perp^2 - (m_e V_{e\perp}^0)^2]\,. \quad (3.7.44)$$

Substituting Eq. (3.7.44) into Eq. (3.7.30), and assuming $k_z = 0$ for simplicity, reduces the dispersion relation to

$$0 = 1 - \frac{\overline{\omega}_{pe}^2}{k_\perp^2}\sum_{n=-\infty}^{\infty}\frac{n(\omega_e^+ - \omega_e^-)}{\omega - \ell\omega_e - n(\omega_e^+ - \omega_e^-)}$$

$$\times\,\frac{1}{V_{e\perp}^0}\,\frac{d}{dV_{e\perp}^0}J_n^2\left(\frac{k_\perp V_{e\perp}^0}{\omega_e^+ - \omega_e^-}\right). \quad (3.7.45)$$

The neutral plasma analog of Eq. (3.7.45) has been studied by Crawford and Tataronis,[155] who find a density threshold for instability. The instability condition for a pure electron plasma can be expressed as $\overline{\omega}_{pe}^2/(\omega_e^+ - \omega_e^-)^2 > 6.62$ or, equivalently, as

$$\frac{2\overline{\omega}_{pe}^2}{\Omega_e^2} > 0.92\,. \quad (3.7.46)$$

Note that the density threshold in Eq. (3.7.46) is slightly below the maximum density limit for existence of the equilibrium [$2\overline{\omega}_{pe}^2/\Omega_e^2 = 1$ at Brillouin flow]. When Eq. (3.7.46) is satisfied, and $2\overline{\omega}_{pe}^2/\Omega_e^2 < 1$, Eq. (3.7.45) has unstable solutions with $\mathrm{Im}\,\omega = \omega_i \approx 0(\omega_e^+ - \omega_e^-)$.

D. Two-Stream Instability

It is evident from Eq. (3.7.30) that momentum-space instabilities associated with the p_z-dependence of $f_e^0(\mathbf{p}^2, p_z)$ may also exist. As an example, consider the equilibrium distribution function

$$f_e^0(\mathbf{p}^2, p_z) = G_e^0(p_\perp^2)F_e^0(p_z)\,, \tag{3.7.47}$$

where $p_\perp^2 = \mathbf{p}^2 - p_z^2$, $2\pi \int_0^\infty dp_\perp\, p_\perp G_e^0(p_\perp^2) = 1$, and

$$F_e^0(p_z) = \frac{\bar{n}_e}{2}\,[\delta(p_z - m_e V_0) + \delta(p_z + m_e V_0)]\ . \tag{3.7.48}$$

Note that the p_z-dependence of $F_e^0(p_z)$ corresponds to two equidensity, counter-streaming electron components, as may be the case when incoming electrons reflect from a magnetic mirror. When Eqs. (3.7.47) and (3.7.48) are substituted into Eq. (3.7.30), it is straightforward to analyze the resulting dispersion relation in the limiting case where†

$$\frac{k_\perp^2 v_{th\perp}^2}{(\omega_e^+ - \omega_e^-)^2} \ll 1\,. \tag{3.7.49}$$

but still $k_\perp^2 R_p^2 \gg 1$ [Eq. (3.7.7)]. In Eq. (3.7.49), $v_{th\perp}^2 = 2\pi \int_0^\infty dp_\perp\, p_\perp (p_\perp/m_e)^2 \times G_e^0(p_\perp^2)$ is the characteristic thermal speed-squared that is associated with the electron motion perpendicular to $B_0\hat{\mathbf{e}}_z$. Since $J_n^2(x) \simeq (x/2)^{2n}$ for $|x| \ll 1$ and $|n| \geq 1$, it is valid to retain only the $n = 0$ term in Eq. (3.7.30) and to approximate $J_0^2[k_\perp v_\perp/(\omega_e^+ - \omega_e^-)] \simeq 1$. For $\ell = 0$ and $k_\perp^2 \ll k_z^2$, the dispersion relation assumes the familiar two-stream form:

$$1 = \frac{\bar{\omega}_{pe}^2/2}{(\omega - k_z V_0)^2} + \frac{\bar{\omega}_{pe}^2/2}{(\omega + k_z V_0)^2}\,. \tag{3.7.50}$$

The solutions to Eq. (3.7.50) are

$$\omega^2 = \frac{1}{2}\,[2k_z^2 V_0^2 + \bar{\omega}_{pe}^2 \pm \bar{\omega}_{pe}(\bar{\omega}_{pe}^2 + 8k_z^2 V_0^2)^{1/2}]\ . \tag{3.7.51}$$

---

†For a pure electron plasma the *thermal* Larmor radius is

$$R_L = [v_{th\perp}^2/(\omega_e^+ - \omega_e^-)^2]^{1/2}\,.$$

Therefore Eqs. (3.7.49) and (3.7.7) require that the transverse wavelength be long in comparison with a thermal Larmor radius ($k_\perp^{-2} \gg R_L^2$), but short in comparison with the column radius ($k_\perp^{-2} \ll R_p^2$).

The lower sign in Eq. (3.7.51) yields one unstable root provided $k_z^2 V_0^2 < \bar{\omega}_{pe}^2$. The maximum growth rate,

$$[\omega_i]_{\max} = [\mathrm{Im}\ \omega]_{\max} = \bar{\omega}_{pe}/2\sqrt{2}\,,$$

occurs for $k_z^2 V_0^2 = (3/8)\,\bar{\omega}_{pe}^2$.

3.7.4 Dispersion Relation for Transverse Electromagnetic Waves

For a nonneutral plasma, an analysis of the linearized Vlasov-Maxwell equations that includes electromagnetic perturbations with arbitrary polarization is somewhat tedious. It is adequate for present purposes to consider the appropriate dispersion relations for two configurations in which the polarization is *purely transverse*. These are illustrated in Figs. 3.7.3 and 3.7.4. As in Sections 3.7.1 - 3.7.3, it is assumed that perturbations are about constant-density rigid-rotor equilibria with $\omega_e \simeq \omega_e^+$ or $\omega_e \simeq \omega_e^-$, and an equilibrium distribution function of the form $f_e^0(H - \omega_e P_\theta, p_z)$. It is also assumed that the perturbations are localized to the column interior ($r < R_p$), with characteristic perpendicular wavelength $|k_\perp|^{-1}$ small in comparison with the column radius R_p [Eq. (3.7.7)].

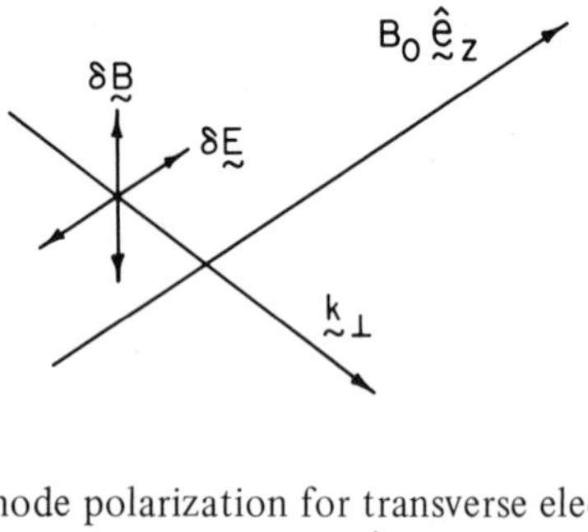

Fig. 3.7.3 Ordinary-mode polarization for transverse electromagnetic waves propagating perpendicular to $B_0\hat{\mathbf{e}}_z$. The perturbed field amplitudes are oriented with $\delta\mathbf{B}\cdot\hat{\mathbf{e}}_z = 0$, $\delta\mathbf{E}$ parallel to $\hat{\mathbf{e}}_z$, and $\delta\mathbf{E}\cdot\delta\mathbf{B} = 0$.

The perturbed field configuration illustrated in Fig. 3.7.3 corresponds to transverse electromagnetic waves propagating perpendicular to $B_0\hat{\mathbf{e}}_z$ with $k_\perp \neq 0$, $k_z = 0$, $\delta\mathbf{E} \parallel \hat{\mathbf{e}}_z$, and $\delta\mathbf{B} \perp \hat{\mathbf{e}}_z$. For a pure electron plasma, the dispersion relation for this configuration can be expressed as[77]

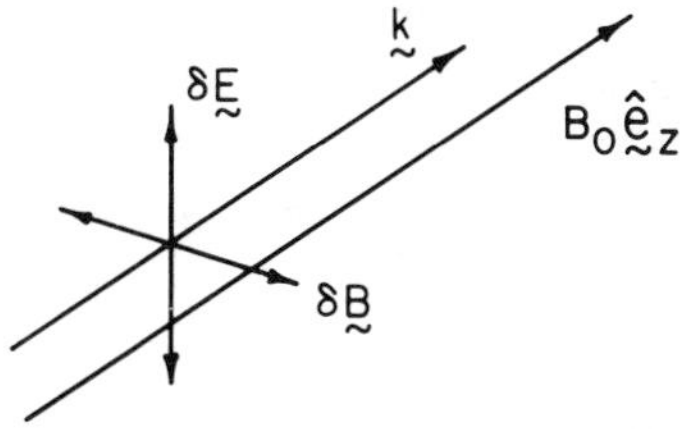

Fig. 3.7.4 Transverse electromagnetic waves propagating parallel to $B_0 \hat{\mathbf{e}}_z$. The perturbed field amplitudes are oriented with $\delta \mathbf{B} \cdot \hat{\mathbf{e}}_z = 0 = \delta \mathbf{E} \cdot \hat{\mathbf{e}}_z$ and $\delta \mathbf{E} \cdot \delta \mathbf{B} = 0$.

$$0 = \omega^2 - c^2 k_\perp^2 - \overline{\omega}_{pe}^2 + \frac{4\pi e^2}{m_e} \sum_{n=-\infty}^{\infty} \frac{n(\omega_e^+ - \omega_e^-)}{\omega - \ell\omega_e - n(\omega_e^+ - \omega_e^-)}$$

$$\times \int d^3p \, J_n^2 \left(\frac{k_\perp v_\perp}{\omega_e^+ - \omega_e^-} \right) \frac{v_z^2}{v_\perp} \frac{\partial}{\partial v_\perp} f_e^0(\mathbf{p}^2, p_z) \,, \tag{3.7.52}$$

where the notation is the same as in Section 3.7.2. If the plasma is cold parallel to the magnetic field, $\int_{-\infty}^{\infty} dp_z v_z^2 f_e^0 = 0$, then Eq. (3.7.52) reduces to the familiar result†

$$\omega^2 = c^2 k_\perp^2 + \overline{\omega}_{pe}^2 \,. \tag{3.7.53}$$

If the plasma is warm and the equilibrium distribution function $f_e^0(\mathbf{p}^2, p_z)$ is an isotropic Maxwellian [Eq. (3.7.37)], the solutions to Eq. (3.7.52) satisfy $\text{Im}\,\omega = 0$, and exhibit an intricate harmonic structure for $\omega - \ell\omega_e$ close to $n(\omega_e^+ - \omega_e^-)$, $n = \pm 1, \pm 2, \pm 3, \cdots$. However, if the equilibrium distribution function is anisotropic, the possibility of electromagnetic instability ($\text{Im}\,\omega > 0$) exists. For example, if $f_e^0(\mathbf{p}^2, p_z)$ is bi-Maxwellian,

$$f_e^0(\mathbf{p}^2, p_z) = \frac{\overline{n}_e}{(2\pi m_e \Theta_{e\perp})(2\pi m_e \Theta_{ez})^{1/2}} \exp\left(-\frac{p_\perp^2}{2m_e\Theta_{e\perp}} - \frac{p_z^2}{2m_e\Theta_{ez}} \right), \tag{3.7.54}$$

† The mode in Eq. (3.7.53) can serve as a useful density diagnostic for nonneutral as well as neutral plasmas.

with $\Theta_{ez} > \Theta_{e\perp}$, then by analogy with the neutral plasma case [156] [157] Eq. (3.7.52) has unstable solutions provided $[\bar{\omega}_{pe}^2/(\omega_e^+ - \omega_e^-)^2]\,(\Theta_{ez}/m_e c^2)$ exceeds a certain threshold value. For $\Theta_{e\perp} \ll \Theta_{ez}$, the condition for instability can be expressed as

$$\frac{\bar{\omega}_{pe}^2}{(\omega_e^+ - \omega_e^-)^2}\,\frac{\Theta_{ez}}{m_e c^2} > 1\,. \tag{3.7.55}$$

Since $\Theta_{ez} \ll m_e c^2$ has been assumed (nonrelativistic assumption), and $(\omega_e^+ - \omega_e^-)^2 = \Omega_e^2(1 - 2\bar{\omega}_{pe}^2/\Omega_e^2)$, Eq. (3.7.55) requires densities very close to Brillouin flow $(2\bar{\omega}_{pe}^2/\Omega_e^2 = 1)$. When Eq. (3.7.55) is satisfied, and $\Theta_{e\perp} \ll \Theta_{ez}$, Eq. (3.7.52) has unstable solutions with $\mathrm{Im}\,\omega = \omega_i \approx 0(\omega_e^+ - \omega_e^-)$.[157]

The perturbed field configuration illustrated in Fig. 3.7.4 corresponds to transverse electromagnetic waves propagating parallel to $B_0\hat{\mathbf{e}}_z$ with $k_z \neq 0$, $k_\perp \simeq 0$,† $\delta\mathbf{E} \perp \hat{\mathbf{e}}_z$, and $\delta\mathbf{B} \perp \hat{\mathbf{e}}_z$. For a pure electron plasma rotating in the *slow* $(\omega_e \simeq \omega_e^-)$ rotational mode, the dispersion relation‡ for this configuration can be expressed as[77]

$$0 = D_-(k_z, \omega) = \omega^2 - c^2 k_z^2 + \frac{4\pi e^2}{m_e}\int d^3p\,(v_\perp/2)$$

$$\times\,\frac{[k_z v_\perp(\partial/\partial v_z) + (\omega - k_z v_z)(\partial/\partial v_\perp)]\,f_e^0(\mathbf{p}^2, p_z)}{\omega - k_z v_z \mp \omega_e^+}\,. \tag{3.7.56}$$

where the notation is the same as in Section 3.7.2. The upper (−) and lower (+) signs in Eq. (3.7.56) correspond to waves with right-hand and left-hand circular polarization, respectively. The dispersion relation in Eq. (2.7.56) is identical in form with the corresponding result for a netural plasma[158] if the replacements $\Omega_e \to \omega_e^+$ and $m_i \to \infty$ are made in the neutral plasma dispersion relation. As in the neutral plasma case, if there is an anisotropy in kinetic energy[159, 160] with $\int d^3p\,p_\perp^2 f_e^0(\mathbf{p}^2, p_z)$ exceeding $\int d^3p\,p_z^2 f_e^0(\mathbf{p}^2, p_z)$ by a sufficient amount, Eq. (3.7.66) supports unstable solutions with $\mathrm{Im}\,\omega > 0$. Examples of unstable equilibria include the loss-cone distribution in Eq. (3.7.44) and the bi-Maxwellian distribution in Eq. (3.7.54) for $\Theta_{e\perp} > \Theta_{ez}$. The corresponding growth rates for a pure electron plasma can be written down by direct analogy with the neutral plasma results.

---

†The limit $k_\perp \to 0$ is approximate since $k_\perp^2 R_p^2 \ll 1$ is required [Eq. (3.7.7)].

---

‡For a pure electron plasma rotating in the *fast* $(\omega_e \simeq \omega_e^+)$ rotational mode the dispersion relation is identical in form to Eq. (3.7.56) with the denominator, $\omega - k_z v_z \mp \omega_e^+$, replaced by $\omega - k_z v_z \mp \omega_e^-$.

It is interesting to note the effect of no neutralizing ion background on the low-frequency long-wavelength modes obtained from Eq. (3.7.56). For $\omega \to 0$ and $k_z \to 0$ [i.e., $|\omega/\omega_e^+| \ll 1$ and $k_z^2 v_{th}^2/\omega_+^2 \ll 1$], Eq. (3.7.56) can be approximated by

$$\omega^2 - c^2 k_z^2 \pm \overline{\omega}_{pe}^2 \omega / \omega_e^+ = 0 . \qquad (3.7.57)$$

As $k_z \to 0$, Eq. (3.7.57) gives $\omega \simeq \pm(c^2 k_z^2/\overline{\omega}_{pe}^2)\omega_e^+$. In contrast to the neutral plasma case, where $\omega^2 = k_z^2 V_A^2$ at low frequencies [158] (here V_A is the Alfvén velocity), the electron whistler mode persists in a pure electron plasma down to zero frequency, and the mode is dispersive ($\omega \sim k_z^2$) as $k_z \to 0$.

REFERENCES

1. J. C. Slater, *Microwave Electronics* (Dover Publications, New York, 1969).
2. L. Tonks and I. Langmuir, "Oscillations in Ionized Gases," *Phys. Rev.* **33**, 195 (1929).
3. W. W. Rigrod and J. A. Lewis, "Wave Propagation Along a Magnetically-Focussed Electron Beam," *Bell System Tech. J.* **33**, 399 (1954).
4. G. R. Brewer, "Some Effects of Magnetic Field Strength on Space-Charge Wave Propagation," *Proc. IRE* **44**, 896 (1956).
5. J. Labus, "Space-Charge Waves Along Magnetically Focussed Electron Beam," *Proc. IRE* **45**, 854 (1957).
6. W. W. Rigrod, "Space-Charge Wave Harmonics and Noise Propagating in Rotating Electron Beams," *Bell System Tech. J.* **38**, 119 (1959)
7. A. W. Trivelpiece and R. W. Gould, "Plasma Waves in Cylindrical Plasma Columns," *J. Appl. Phys.* **30**, 1784 (1959).
8. H. Pötzl, "Types of Waves in Magnetically Focussed Electron Beams," *Arch. Elek. Übertragung* **19**, 367 (1965).
9. R. C. Davidson, "Electrostatic Shielding of a Test Charge in a Nonneutral Plasma," *J. Plasma Phys.* **6**, 229 (1971).
10. D. Keefe, G. R. Lambertson, L. J. Laslett, W. A. Perkins, J. M. Peterson, A. M. Sessler, R. W. Allison, Jr., W. W. Chupp, A. U. Luccio, and J. B. Rechen, "Experiments on Forming Intense Rings of Electrons Suitable for Acceleration of Ions," *Phys. Rev. Letters* **22**, 558 (1969).
11. D. Keefe, "The Electron Ring Accelerator," *IEEE Trans. Nucl. Sci.* **NS-16**, 25 (1969).
12. D. Keefe, "Research on the Electron Ring Accelerator," *Particle Accelerators* **1**, 1 (1970).
13. G. R. Lambertson, D. Keefe, L. J. Laslett, W. A. Perkins, J. M. Peterson, and J. B. Rechen, "Recent Experiments on Forming Electron Rings at Berkeley," *IEEE Trans. Nucl. Sci.* **NS-18**, 501 (1971).
14. D. Keefe, W. W. Chupp, A. A. Garren, G. R. Lambertson, L. J. Laslett, A. U. Luccio, W. A. Perkins, J. M. Peterson, J. B. Rechen, and A. M. Sessler,

"Experiments on Forming, Compressing and Extracting Electron Rings for the Collective Acceleration of Ions," *Nucl. Instr. Methods* **93**, 541 (1971).

15. V. I. Veksler, V. P. Sarantsev, et al., "Collective Linear Acceleration of Ions," in *Proceedings 6th International Conference on Accelerators, Cambridge, Mass., 1967* (Cambridge Electron Accelerator Report No. CEAL-2000, 1967), p. 289.
16. V. I. Veksler, V. P. Sarantsev, A. G. Bonch-Osmolovskii, G. V. Dolbilov, G. A. Ivanov, I. N. Ivanov, M. L. Iovonich, I. V. Kozhuhov, A. B. Kuznetsov, V. G. Mahankov, E. A. Perelstein, V. P. Rashevskii, K. A. Reshetnikova, N. B. Rubin, S. B. Rubin, P. I. Ryltsev, and O. I. Yarkovov, "Collective Linear Acceleration of Ions," *Atomnaya Energiya (USSR)* **24**, 317 (1968).
17. V. P. Sarantsev, "Collective Method of Proton Acceleration," *IEEE Trans. Nucl. Sci.* **NS-16**, 15 (1969).
18. V. P. Sarantsev, "Status Report on the Collective Linear Accelerator at Dubna," in *Proceedings 8th International Conference on High-Energy Accelerators, 1971* (CERN Scientific Information Service, Geneva, 1971), p. 391.
19. C. Andelfinger, W. Herrmann, A. Schluter, U. Schumacher, and M. Ulrich, "Measurements of Electron Ring Compression in the Garching ERA," *IEEE Trans. Nucl. Sci.* **NS-18**, 505 (1971).
20. R. E. Berg, H. Kim, M. P. Reiser, and G. T. Zorn, "Possibility of Forming a Compressed Electron Ring in a Static Magnetic Field," *Phys. Rev. Letters,* **22**, 419 (1969).
21. M. Reiser, "The University of Maryland Electron Ring Accelerator Concept," *IEEE Trans. Nucl. Sci.* **NS-18**, 460 (1971).
22. M. J. Rhee, G. T. Zorn, R. C. Placious, and J. H. Sparrow, "Studies of Electron Beams from a Febetron 70," *IEEE Trans. Nucl. Sci.* **NS-18**, 468 (1971).
23. M. Reiser, "Ion Loading and Acceleration in a Static-Field ERA," *IEEE Trans. Nucl. Sci.* **NS-19**, 280 (1972).
24. M. Reiser, "Status Report on the University of Maryland Electron Ring Accelerator Project," *IEEE Trans. Nucl. Sci.* **NS-20**, 310 (1973).
25. R. C. Davidson and J. D. Lawson, "Self-Consistent Vlasov Description of Relativistic Electron Rings," *Particle Accelerators* **4**, 1 (1972).
26. J. D. Lawson, "Collective and Coherent Methods of Particle Acceleration," *Particle Accelerators* **3**, 21 (1972).
27. D. Keefe, "Collective-Effect Accelerators," *Sci. Am.* **226**, 22 (1972).
28. H. Alfvén and P. Wernholm, "A New Type of Accelerator," *Arkiv Fysik* **5**, 175 (1952).
29. V. I. Veksler, "Coherent Principle of Acceleration of Charged Particles," in *Proc. CERN Sym. on High-Energy Accelerators and Pion Physics, Geneva, 1956* (CERN Scientific Information Service, Geneva, 1956), Vol. 1, p. 80.

30. G. J. Budker, "Relativistic Stabilized Electron Beam," in *Proc. CERN Sym. on High-Energy Accelerators and Pion Physics, Geneva, 1956* (CERN Scientific Information Service, Geneva, 1956), Vol. 1, p. 68.

31. Ya. B. Fainberg, "The Use of Plasma Waveguides as Accelerating Structures in Linear Accelerators," in *Proc. CERN Sym. on High-Energy Accelerators and Pion Physics, Geneva, 1956* (CERN Scientific Information Service, Geneva, 1956), Vol. 1, p. 84.

32. S. E. Graybill and S. V. Nablo, "Observations of Magnetically Self-Focussing Electron Streams," *Appl. Phys. Letters* **8**, 18 (1966).

33. W. T. Link, "Electron Beams from 10^{11}–10^{12} Watt Pulsed Accelerators," *IEEE Trans. Nucl. Sci.* **14**, 777 (1967).

34. T. G. Roberts and W. H. Bennett, "The Pinch Effect in Pulsed Streams at Relativistic Energies," *Plasma Phys.* **10**, 381 (1968).

35. G. Yonas and P. Spense, "Experimental Investigations of High ν/γ Beam Transport," in *Rec. 10th Symp. on Electron, Ion and Laser Beam Technology* (San Francisco Press, San Francisco, Calif., 1969), p. 143.

36. J. R. Uglum, W. H. McNeill, and S. E. Graybill, "Beam Characteristics of Intense Relativistic Electron Accelerators," in *Rec. 10th Symp. on Electron, Ion and Laser Beam Technology* (San Francisco Press, San Francisco, Calif., 1969), p. 155.

37. L. S. Levine, I. M. Vitkovitsky, D. A. Hammer, and M. L. Andrews, "Propagation of an Intense Relativistic Electron Beam Through a Plasma Background," *J. Appl. Phys.* **42**, 1863 (1971).

38. L. P. Bradley, T. H. Martin, K. R. Prestwich, J. E. Boers, and D. L. Johnson, "Characteristics of Cool, High ν/γ Electron Beam Propagation," in *Rec. 11th Symp. on Electron, Ion and Laser Beam Technology* (San Francisco Press, San Francisco, Calif., 1971), p. 553.

39. J. Benford and B. Ecker, "Transport of Intense Relativistic Electron Beams in a Z Pinch," *Phys. Rev. Letters* **26**, 1160 (1971).

40. J. Block, J. Burton, J. M. Frame, D. Hammer, A. C. Kolb, L. S. Levine, W. H. Lupton, W. F. Oliphant, J. D. Shipman, Jr., and I. M. Vitkovitsky, "NRL Relativistic Electron Beam Program," in *Rec. 11th Symp. on Electron, Ion and Laser Beam Technology* (San Francisco Press, San Francisco, Calif., 1971), p. 513.

41. S. E. Graybill, "Dynamics of Pulsed High Current Relativistic Electron Beams," *IEEE Trans. Nucl. Sci.* **NS-18**, 438 (1971).

42. M. Friedman and D. A. Hammer, "Catastrophic Disruption of the Flow of a Magnetically Confined Relativistic Electron Beam," *Appl. Phys. Letters* **21**, 174 (1972).

43. D. A. Hammer, W. F. Oliphant, I. M. Vitkovitsky, and V. Fargo, "Interaction of Accelerating High-Current Electron Beams with External Magnetic Fields," *J. Appl. Phys.* **43**, 58 (1972).

44. W. H. Bennett, "Magnetically Self-Focussing Streams," *Phys. Rev.* **45**, 890 (1934).

45. M. Friedman and M. Herndon, "Microwave Emission Produced by the Interaction of an Intense Relativistic Electron Beam with a Spatially Modulated Magnetic Field," *Phys. Rev. Letters* **28**, 210 (1972).

46. J. Nation, "On the Coupling of a High Current Relativistic Electron Beam to a Slow Wave Structure," *Appl. Phys. Letters* **17**, 491 (1970).

47. F. Winterberg, "The Possibility of Producing a Dense Thermonuclear Plasma by an Intense Field Emission Discharge," *Phys. Rev.* **174**, 212 (1968).

48. M. Friedman, "Passage of an Intense Relativistic Electron Beam Through a Cusped Magnetic Field," *Phys. Rev. Letters* **24**, 1098 (1970).

49. M. Friedman, "Another Approach to the Injection of Relativistic Electrons into an Astron-like Device," *Phys. Rev. Letters* **25**, 567 (1970).

50. M. L. Andrews, H. Davitian, H. H. Fleischmann, B. Kusse, R. E. Kribel, and J. A. Nation, "Generation of Astron-Type E Layers Using Very High-Current Electron Beams," *Phys. Rev. Letters* **27**, 1428 (1971).

51. M. L. Andrews, H. Davitian, H. H. Fleischmann, R. E. Kribel, V. R. Cusse, J. A. Nation, R. Lee, R. V. Lovelace, and R. N. Sudan, "Application of Intense Relativistic Electron Beams to Astron-Type Experiments," in *Plasma Physics and Controlled Nuclear Fusion Research* (International Atomic Energy Agency, Vienna, 1971), Vol. 1, p. 169.

52. S. D. Putnam, "Model of Energetic Ion Production by Intense Electron Beams," *Phys. Rev. Letters* **25**, 1129 (1970).

53. S. E. Graybill and J. R. Uglum, "Observation of Heavy Ions from a Beam Generated Plasma," *J. Appl. Phys.* **41**, 236 (1970).

54. S. Putnam, "Ion Acceleration with Intense Linear Electron Beams," *IEEE Trans. Nucl. Sci.* **NS-18**, 496 (1971).

55. S. E. Graybill, W. H. McNeill, and J. R. Uglam, "Acceleration of Ions by Relativistic Electron-Beam-Formed Plasmas," in *Rec. 11th Symp. on Electron, Ion and Laser Beam Technology* (San Francisco Press, San Francisco, Calif., 1971), p. 577.

56. A. T. Altyntsev, B. N. Breyzman, A. G. Eskov, O. A. Zolotivskii, V. I. Koroteev, R. Jurtmellaer, V. L. Nasalov, D. D. Ryutov, and V. N. N. Semenov, in *Plasma Physics and Controlled Nuclear Fusion Research* (International Atomic Energy Agency, Vienna, 1971), Vol. 2, p. 309.

57. C. A. Kapetanakos and D. A. Hammer, "Plasma Heating by an Intense Relativistic Electron Beam," *Appl. Phys. Letters* **23**, 17 (1973).

58. A. A. Ivanov and L. I. Rudakov, "Intense Relativistic Electron Beam in a Plasma," *Sov. Phys.–JETP* **31**, 715 (1970).

59. L. S. Bogdankevich and A. A. Rukhadze, "Stability of Relativistic Electron Beams in a Plasma and the Problem of Critical Currents," *Sov. Phys.–Usp.* **14**, 163 (1971), and references therein.

60. Ya. B. Fainberg, V. D. Shapiro, and V. I. Shevchenko, "Nonlinear Theory of Interaction Between a Monochromatic Beam of Relativistic Electrons and a Plasma," *Sov. Phys.–JETP* **3**, 528 (1970).

61. L. I. Rudakov, "Collective Slowing Down of an Intense Beam of Relativistic Electrons in Dense Plasma Target," *Sov. Phys.–JETP* **32**, 1134 (1971).

62. R. V. Lovelace and R. N. Sudan, "Plasma Heating by High-Current Relativistic Electron Beams," *Phys. Rev. Letters* **27**, 1256 (1971).

63. G. Benford, "Electron Beam Filamentation in Strong Magnetic Fields," *Phys. Rev. Letters* **28**, 1242 (1972).

64. K. R. Chu and N. Rostoker, "Current Neutralization and Energy Transfer Processes of a Hollow Rotating Relativistic Electron Beam in a Magnetized Plasma," *Phys. Fluids* **16**, 1472 (1973).

65. G.S. Janes, R. H. Levy, H. A. Bethe, and B. T. Feld, "New Type of Accelerator for Heavy Ions," *Phys. Rev.* **145**, 925 (1966).

66. J. D. Daugherty, L. Grodzins, G. S. Janes, and R. H. Levy, "New Source of Highly Stripped Heavy Ions," *Phys. Rev. Letters* **20**, 369 (1968).

67. J. D. Daugherty, J. E. Eninger, and G. S. Janes, "Experiments on the Injection and Containment of Electron Clouds in a Toroidal Apparatus," *Phys. Fluids* **12**, 2677 (1969).

68. R. H. Levy, "Diocotron Instability in Cylindrical Plasma," *Phys. Fluids* **8**, 1288 (1965).

69. J. D. Daugherty and R. H. Levy, "Equilibrium of Electron Clouds in Toroidal Magnetic Fields," *Phys. Fluids* **10**, 155 (1967).

70. R. H. Levy, J. D. Daugherty, and O. Buneman, "Ion Resonance Instability in Grossly Nonneutral Plasmas," *Phys. Fluids* **12**, 2616 (1969).

71. R. E. Pechacek, C. A. Kapetanakos, and A. W. Trivelpiece, "Trapping of a 0.5 MeV Electron Ring in a 15 kG Pulsed Magnetic Mirror Field," *Phys. Rev. Letters* **21**, 1436 (1968).

72. C. A. Kapetanakos, R. E. Pechacek, D. M. Spero, and A. W. Trivelpiece, "Trapping and Confinement of Nonneutral Hot Electron Clouds in a Magnetic Mirror," *Phys. Fluids* **14**, 1555 (1971).

73. C. P. DeNeef, R. E. Pechacek, and A. W. Trivelpiece, "Comparison of the Hot Electron Plasmas Produced Using Two Different Plasma Sources in a Magnetic Mirror Compression Experiment," *Phys. Fluids* **16**, 509 (1973).

74. A. W. Trivelpiece, "Nonneutral Plasmas," *Commun. Plasma Phys. Controlled Fusion* **1**, 57 (1972).

75. R. C. Davidson, A. Drobot, and C. A. Kapetanakos, "Equilibrium and Stability of Mirror-Confined Nonneutral Plasmas," *Phys. Fluids* **16**, 2199 (1973).

76. R. C. Davidson and N. A. Krall, "Vlasov Description of an Electron Gas in a Magnetic Field," *Phys. Rev. Letters* **22**, 833 (1969).

77. R. C. Davidson and N. A. Krall, "Vlasov Equilibria and Stability of an Electron Gas," *Phys. Fluids* **13**, 1543 (1970).

78. B. L. Bogema and R. C. Davidson, "Rotor Equilibria of Nonneutral Plasmas," *Phys. Fluids* **13**, 2772 (1970).

79. R. C. Davidson and N. A. Krall, "A Characteristic Instability of an Electron Gas of Uniform Density," *Phys. Letters* **32A**, 187 (1970).

80. B. L. Bogema and R. C. Davidson, "Two-Rotating-Stream Instability in a Nonneutral Plasma," *Phys. Fluids* **14**, 1456 (1971).

81. A. Theiss, "An Investigation of the Equilibrium Properties and Symmetric Wave Propagation Properties of Rigid-Rotor Nonneutral Plasmas," Ph. D. Thesis, University of Maryland, 1973 (University of Maryland Tech. Rep. No. 74–093).

82. T. H. Stix, "Toroidal Fusion Plasma with Powerful Negative Bias," *Phys. Rev. Letters* **24**, 135 (1970).

83. T. H. Stix, "Some Toroidal Equilibria for Plasma Under Magnetoelectric Confinement," *Phys. Fluids* **14**, 692 (1971).

84. T. H. Stix, "Stability of Cold Plasma Under Magnetoelectric Confinement," *Phys. Fluids* **14**, 702 (1971).

85. T. H. Stix, "Negatively Charged Open-Ended Plasma to Strip and Confine Heavy Ions," *Phys. Rev. Letters* **23**, 1093 (1969).

86. H. Griem, *Plasma Spectroscopy* (McGraw-Hill, New York, 1964).

87. A. A. Vlasov, "On the Kinetic Theory of an Assembly of Particles with Collective Interaction," *J. Phys. (USSR)* **9**, 25 (1945).

88. N. A. Krall and A. W. Trivelpiece, *Principles of Plasma Physics* (McGraw-Hill, New York, 1973).

89. D. C. Montgomery, *Kinetic Theory of the Coulomb Plasma* (Gordon and Breach, New York, 1972).

90. Y. L. Klimontovich, *The Statistical Theory of Nonequilibrium Processes in Plasmas* (MIT Press, Cambridge, Mass., 1967).

91. D. C. Montgomery and D. A. Tidman, *Plasma Kinetic Theory* (McGraw-Hill, New York, 1964).

92. T. H. Stix, *The Theory of Plasma Waves* (McGraw-Hill, New York, 1962).

93. See, for example, Chapters 3–5 of Reference 88.

94. J. D. Lawson, "On the Adiabatic Self-Constriction of an Accelerated Electron Beam Neutralized by Positive Ions," *J. Electron. Control* **3**, 587 (1957).

95. J. D. Lawson, "Perveance and the Bennett Pinch Relation in Partially Neutralized Electron Beams," *J. Electron. Control* **5**, 146 (1958).

96. J. D. Lawson, "On the Classification of Electron Streams," *J. Nucl. Energy, Part C: Plasma Phys.* **1**, 31 (1959)

97. A. W. Trivelpiece, *Slow-Wave Propagation in Plasma Waveguides*, (San Francisco Press, San Francisco, Calif., 1967).

98. C. C. MacFarlane and H. G. Hay, "Wave Propagation in a Slipping Stream of Electrons; Small-Amplitude Theory," *Proc. Phys. Soc. (London)* **63B**, 409 (1950).

99. H. F. Webster, "Breakup of Hollow Electron Beams," *J. Appl. Phys.* **26**, 1386 (1955).

100. C. C. Cutler, "Instability in Hollow and Strip Electron Beams," *J. Appl. Phys.* **27**, 1028 (1956).

101. J. R. Pierce, "Instability of Hollow Beams," *IRE Trans. Electron Devices* **ED-3**, 183 (1956).

102. R. L. Kyhl and H. F. Webster, "Breakup of Hollow Cylindrical Electron Beams," *IRE Trans. Electron Devices* **ED-3**, 172 (1956).

103. G. R. Brewer, "Some Characteristics of a Cylindrical Electron Stream in Immersed Flow," *IRE Trans. Electron Devices* **ED-4**, 134 (1957).

104. O. Buneman, "Ribbon Beams," *J. Electron. Control* **3**, 507 (1957).

105. V. K. Neil and W. Heckrotte, "Relation Between Diocotron and Negative-Mass Instabilities," *J. Appl. Phys.* **36**, 2761 (1965).

106. R. H. Levy, "Diocotron Instability in Cylindrical Plasma," *Phys. Fluids* **8**, 1288 (1965).

107. W. Knauer, "Diocotron Instability in Plasmas and Gas Discharges," *J. Appl. Phys.* **37**, 602 (1966).

108. R. H. Levy, "Two New Results in Cylindrical Diocotron Theory," *Phys. Fluids* **11**, 920 (1968).

109. A. Nocentini, H. L. Berk, and R. N. Sudan, "Kinetic Theory of the Diocotron Instability," *J. Plasma Phys.* **2**, 311 (1968).

110. R. J. Briggs, J. D. Daugherty, and R. H. Levy, "Role of Landau Damping in Crossed-Field Electron Beams and Inviscid Shear Flow," *Phys. Fluids* **13**, 421 (1970).

111. C. A. Kapetanakos, D. A. Hammer, C. Striffler, and R. C. Davidson, "Destructive Instabilities in Hollow Intense Relativistic Electron Beams," *Phys. Rev. Letters* **30**, 1303 (1973).

112. L. Tonks, "The Structure of the Astron E-Layer," *Nucl. Fus.* **1**, 273 (1961).

113. D. Pfirsch, "Mikroinstabilitäten vom Spiegeltyp in Inhomogenen Plasmen," *Z. Naturforsch.* **17a**, 861 (1962).

114. K. D. Marx, "Equilibria and Stability of Collisionless Plasmas in Cylindrical Geometry," *Phys. Fluids* **11**, 357 (1968).

115. B. Marder and H. Weitzner, "A Bifurcation Problem in E-Layer Equilibria," *Plasma Phys.* **12**, 435 (1970).

116. S. Fisher, "Reversed Field Equilibria in Axially Symmetric Electron Current Layers," *Phys. Fluids* **14**, 962 (1971).

117. D. V. Anderson, J. Killeen, and M. E. Rensink, "Computation of E-Layer and Plasma Equilibria in Astron," *Phys. Fluids* **15**, 351 (1972).

118. M. E. Rensink, "E-Layer Compression by a Toroidal Magnetic Field," *Phys. Fluids* **15**, 2391 (1972).

119. M. E. Rensink, "Thin E-Layer Equilibria," *Phys. Fluids* **16**, 443 (1973).

120. N. C. Christofilos, "Astron Thermonuclear Reactor," in *Proceedings 2nd International Conference on the Peaceful Uses of Atomic Energy* (United Nations, Geneva, 1958), Vol. 32, p. 279.

121. N. C. Christofilos, W. C. Condit, Jr., T. J. Fessenden, R. E. Hester, S. Humphries, G. D. Porter, V. W. Stallard and P. V. Weiss, "Trapping Experiments in the Astron," in *Plasma Physics and Controlled Nuclear Fusion Research* (International Atomic Energy Agency, Vienna, 1971), Vol. 1, p. 119.

122. D. A. Hammer and N. Rostoker, "Propagation of High Current Relativistic Electron Beams," *Phys. Fluids* **13**, 1831 (1970).

123. G. Benford, D. L. Book, and R. N. Sudan, "Relativistic Beam Equilibria with Back Currents," *Phys. Fluids* **13**, 2621 (1970).

124. M. E. Rensink, "Self-Consistent Relativistic Beam Equilibria," *Phys. Fluids* **14**, 2241 (1971).

125. J. P. Boris and R. Lee, "Computational Studies of Relativistic Electron Beams," in *Rec. 11th Symp. Electron, Ion and Laser Beam Technology* (San Francisco Press, San Francisco, Calif., 1971), p. 535.

126. G. Benford and D. L. Book, "Relativistic Beam Equilibria," in *Advances in Plasma Physics* (Eds. A. Simon and W. B. Thompson, John Wiley and Sons, New York, 1971), Vol. 4, p. 125.

127. R. C. Davidson and C. D. Striffler, "Vlasov Equilibria for Intense Hollow Relativistic Electron Beams," *J. Plasma Phys.* in press (1974).

128. P. Grateau, "Détermination de l'état Stationnaire d'un Faisceau Annulaire d'electrons Monoenergetiques," *Plasma Phys.* **11**, 1001 (1969).

129. W. H. Kegel, "Some Properties of Relativistic Plasma Rings," *Plasma Phys.* **12**, 105 (1970).

130. E. Ott, "Toroidal Equilibria of Electrically Unneutralized Intense Relativistic Electron Beams," *Plasma Phys.* **13**, 529 (1971).

131. G. Schmidt, "Self-Consistent Field Theory of Relativistic Electron Rings," *Phys. Rev. Letters* **26**, 952 (1971).

132. R. C. Davidson and S. Mahajan, "A Relativistic Electron Ring Equilibrium with Thermal Energy Spread," *Particle Accelerators* **4**, 53 (1972).

133. L. Brillouin, "A Theorem of Larmor and Its Importance for Electrons in Magnetic Fields," *Phys. Rev.* **67**, 260 (1945).

134. A. I. Samuel, "On the Theory of Axially Symmetric Electron Beams in an Axial Magnetic Field," *Proc. IRE* **37**, 1252 (1949).

135. L. D. Landau, "On the Vibrations of the Electronic Plasma," *J. Phys. (USSR)* **10**, 25 (1946).

136. See, for example, Section 10.1 of Reference 91.

137. See, for example, Section 7.7 of Reference 88.

138. B. L. Bogema, "Equilibrium and Stability Theory of Nonneutral Plasmas," Ph. D. Thesis, University of Maryland, 1971 (University of Maryland Tech. Rep. No. 72–037).

139. B. Hornady, "Analysis of Two Nonneutral Plasma Equilibria; Measurement of the Trapped Electron Energy Distribution," Ph. D. Thesis, University of Maryland, 1973 (University of Maryland Tech. Rep. No. 74–027).

140. L. M. Linson, "Electrostatic Electron Oscillations in Cylindrical Nonneutral Plasmas," *Phys. Fluids* **14**, 805 (1971).

141. See, for example, Section 4 of "Plasma Oscillations (I)," by I. B. Bernstein and S. K. Trehan, *Nucl. Fus.* **1**, 3 (1960).

142. H. Alfvén, "On the Motion of Cosmic Rays in Interstellar Space," *Phys. Rev.* **55**, 425 (1939).

143. J. L. Hieronymus, "Equilibria and Stability of Magnetically Confined Electron Clouds," Ph.D. Thesis, Cornell University, 1971.

144. J. D. Lawson, "The Dynamics of High Current Electron Ring Beams in a Time Varying Magnetic Field with Axial Symmetry," *Phys. Letters* **29A**, 344 (1969).

145. I. M. Kapchinskij and V. V. Vladimirskij, "Limitations of Proton Beam Current in a Strong Focussing Linear Accelerator Associated with the Beam Space Charge," in *Proc. 2nd International Conf. on High-Energy Accelerators, 1959* (CERN Scientific Information Service, Geneva, 1959), p. 274.

146. T. R. Walsh, "A Normal Beam with Linear Focusing and Space-Charge Forces," *Plasma Phys.* **5**, 17 (1963).

147. See, for example, Section 12 of "Plasma Oscillations (II)," by I. B. Bernstein, S. K. Trehan, and M. P. H. Weenink, *Nucl. Fus.* **4**, 61 (1964).

148. H. V. Wong, M. L. Sloan, J. R. Thompson, and A. T. Drobot, "Stability of an Unneutralized Rigidly Rotating Electron Beam," *Phys. Fluids* **16**, 902 (1973).

149. C. S. Gardner, "Bound on the Energy Available from a Plasma," *Phys. Fluids* **6**, 839 (1963).

150. See, for example, Section 2 of "Kinetic Theory of Plasma Waves in a Magnetic Field," by D. E. Baldwin, I. B. Bernstein, and M. P. H. Weenink, in *Advances in Plasma Physics* (A. Simon and W. Thompson Eds., John Wiley and Sons, New York, 1969), Vol. 3, p. 1.

151. See, for example, Section 8.5 of Reference 88.

152. See, for example, R. C. Davidson, *Methods in Nonlinear Plasma Theory* (Academic Press, New York, 1972), Chapter 9.

153. I. B. Bernstein, "Waves in a Plasma in a Magnetic Field," *Phys. Rev.* **109** 10 (1958).

154. R. A. Dory, G. E. Guest, and E. G. Harris, "Unstable Electrostatic Plasma Waves Propagating Perpendicular to a Magnetic Field," *Phys. Rev. Letters* **14**, 131 (1965).

155. F. W. Crawford and J. A. Tataronis, "Absolute Instabilities of Perpendicularly Propagating Cyclotron Harmonic Plasma Waves," *J. Appl. Phys.* **36**, 2930 (1965).

156. R. C. Davidson and C. S. Wu, "Ordinary-Mode Electromagnetic Instability in High-β Plasmas," *Phys. Fluids* **13**, 1407 (1970).

157. S. Hamasaki, "Stability of Electromagnetic Waves Propagating Perpendicular to a Uniform Magnetic Induction," *Phys. Fluids* **11**, 1173 (1968).

158. See, for example, Section 10.5 of Reference 91.

159. R. N. Sudan, "Plasma Electromagnetic Instabilities," *Phys. Fluids* **6**, 57 (1963).

160. J. E. Scharer and A. W. Trivelpiece, "Cyclotron Wave Instabilities in a Plasma," *Phys. Fluids* **10**, 591 (1967).

SUPPLEMENTARY REFERENCES

The following references, while not cited directly in the text, are also relevant to the general subject areas of this book.

ION ACCELERATION

"The Acceleration of Heavy Ions", R. S. Livingston, *Particle Accelerators* **1**, 51 (1970).

"Nonlinear Theory of Collective Acceleration of Ions by a Relativistic Electron Beam", V. B. Krasovitskii, *Sov. Phys. JETP* **32**, 98 (1971).

"Trends in the United States for the Acceleration of Heavy Ions", R. S. Livingston and J. A. Martin, *Particle Accelerators* **2**, 189 (1971).

"Neutron Production and Collective Ion Acceleration in a High-Current Diode", L. P. Bradley and G. W. Kuswa, *Phys. Rev. Letters* **29**, 1441 (1972).

"Autoresonant Accelerator Concept", M. L. Sloan and W. E. Drummond, *Phys. Rev. Letters* **31**, 1234 (1973).

RELATIVISTIC ELECTRON BEAMS

"Statistical Mechanics of Relativistic Streams", K. M. Watson, S. A. Bludman and M. N. Rosenbluth, *Phys. Fluids* **3**, 741 (1960); *Phys. Fluids* **3**, 747 (1960).

"Brillouin Flow in Relativistic Beams", D. C. dePach and P. Ulrich, *J. Electron Control* **10**, 139 (1961).

"Kinetic Treatment of the Stability of a Relativistic Particle Beam Passing Through a Plasma", R. C. Mjolsness, *Phys. Fluids* **6**, 1730 (1963).

"Application of the Vlasov Equation to the High Frequency Stability of Finite Relativistic Beam-Plasma Systems", G. Dorman, *J. Appl. Phys.* **37**, 2321 (1966).

"Reverse Current Induced by Injection of a Relativistic Electron Beam Into a Pinched Plasma", J. L. Cox and W. H. Bennett, *Phys. Fluids* **13**, 182 (1970).

"Quasilinear Relaxation of an Ultrarelativistic Electron Beam in a Plasma", B. N. Breizman and D. D. Ryutov, *Sov. Phys. JETP* **33**, 220 (1971).

"Return Current Induced by a Relativistic Beam Propagating in a Magnetized Plasma", R. Lee and R. N. Sudan, *Phys. Fluids* **14**, 1213 (1971).

"Return Current Induced by a Relativistic Electron Beam Propagating into Neutral Gas", D. A. McArthur and J. W. Poukey, *Phys. Rev. Letters* **27**, 1765 (1971).

"One-Dimensional Simulation of Relativistic Streaming Instabilities", C. F. McKee, *Phys. Fluids* **14**, 2164 (1971).

"Finite Beta Equilibria of Relativistic Electron Beams in Toroidal Geometry", E. Ott and R. N. Sudan, *Phys. Fluids* **14**, 1226 (1971).

"One-Dimensional Model of Relativistic Electron Beam Propagation", J. W. Poukey and N. Rostoker, *Plasma Phys.* **13**, 897 (1971).

"Axially Dependent Equilibria for a Relativistic Electron Beam", J. W. Poukey, A. J. Toepfer and J. G. Kelly, *Phys. Rev. Letters* **26**, 1620 (1971).

"Finite-Temperature Relativistic Electron Beam", A. J. Toepfer, *Phys. Rev.* **A3**, 1444 (1971).

"Force-Free Configuration of a High-Intensity Electron Beam", S. Yoshikawa, *Phys. Rev. Letters* **26**, 295 (1971).

"Nonlinear Theory of Dissipative Instability of a Relativistic Beam in a Plasma", V. U. Abramovich and V. I. Shevchenko, *Sov. Phys. JETP* **35**, 730 (1972).

"Estimates of Dense Plasma Heating by Stable Intense Electron Beams", J. Guillory and G. Benford, *Plasma Phys.* **14**, 1131 (1972).

"Theoretical Studies of Intense Relativistic Electron Beam-Plasma Interactions", S. Putman, *Physics International Report PIFR-72-105* (Physics International, San Leandro, Calif., 1972).

"Behavior of a Large-Current Electron Beam in a Dense Gas", L. I. Rudakov, V. P. Smirnov, and A. M. Spektor, *JETP Letters* **15**, 382 (1972).

"Nonlinear Waves in Uncompensated Electron Beams", Va. B. Fainberg, V. D. Shapiro, and V. I. Shevchenko, *Sov. Phys. JETP* **34**, 103 (1972).

"Injection of a Relativistic Electron Beam into a Plasma", A. A. Rukhadze and V. G. Rukhlin, *Sov. Phys. JETP* **34**, 93 (1972).

"Theory of Filamentation in Relativistic Electorn Beams", G. Benford, *Plasma Physics* **15**, 433 (1973).

"Enhanced Microwave Emission due to the Transverse Energy of a Relativistic

Electron Beam", M. Friedman, D. A. Hammer, W. M. Manheimer and P. Sprangle, *Phys. Rev. Letters* 31, 752 (1973).

"Generation of Intense Infrared Radiation from an Electron Beam Propagating through a Rippled Magnetic Field", M. Friedman and M. Herndon, *Appl. Phys. Letters* **22**, 658 (1973).

"Stability of Large-Amplitude Relativistic Plasma Waves", J. Jancarik and V. N. Tsytovich, *Nucl. Fus.* **13**, 807 (1973).

"Macroscopic Equilibria of Relativistic Electron Beams in Plasmas", G. Küppers, A. Salat and H. K. Wimmel, *Plasma Physics* **15**, 441 (1973).

"Electromagnetic Instabilities, Filamentation, and Focusing of Relativistic Electron Beams", R. Lee and M. Lampe, *Phys. Rev. Letters* **31**, 1390 (1973).

"Plasma Heating by an Intense Relativistic Electron Beam", P. A. Miller and G. W. Kuswa, *Phys. Rev. Letters* **30**, 958 (1973).

"Cone Focusing of Intense Relativistic Electron Beams", D. L. Olson, *Phys. Fluids* **16**, 529 (1973).

"Adiabatic Envelope Conservation in Cone Focus Trajectories", C. L. Olson, *Phys. Fluids* **16**, 539 (1973).

"Two-Stream Instability Heating of Plasmas by Relativistic Electron Beams", L. E. Thode and R. N. Sudan, *Phys. Rev. Letters* **30**, 732 (1973).

"Saturation of the Relativistic Beam-Plasma Instability for Arbitrary Density Ratios", A. J. Toepfer and J. W. Poukey, *Phys. Letters* **42A**, 383 (1973).

"Automodulation of an Intense Relativistic Electron Beam", M. Friedman, *Phys. Rev. Letters* **32**, 92 (1974).

"Plasma Heating by Intense Relativistic Electron Beams", G. C. Goldenbaum, W. F. Dove, K. A. Gerber and B. G. Logan, *Phys. Rev. Letters* **32**, 830 (1974).

"Energy Transport Efficiency of an Intense Relativistic Electron Beam Through a Cusped Magnetic Field", C. A. Kapetanakos, *Appl. Phys. Letters* **24**, 112 (1974).

"Theory of Microwave Generation by an Intense Relativistic Electron Beam in a Rippled Magnetic Field", W. M. Manheimer and E. Ott, *Phys. Fluids* **17**, in press (1974).

RELATIVISTIC ELECTRON RINGS AND E-LAYERS

"Longitudinal Instabilities in Intense Relativistic Beams", C. E. Neilsen, A. M. Sessler and K. R. Symon, in *Proc. 2nd International Conf. on High-Energy Accelerators,* 1959 (CERN Scientific Information Service, Geneva, 1959), p. 235.

"Beam Stability in Stacked Orbits", A. A. Kolomenskii and A. N. Lebedev, *J. Nucl. Energy* **C3**, 44 (1961).

"Negative Mass Instability", R. W. Landau and V. K. Neil, *Phys. Fluids* **9**, 2412 (1966).

"Effects of Cold Plasma on the Negative Mass Instability of a Relativistic Electron Layer", Y. Y. Lau and R. J. Briggs, *Phys. Fluids* **14**, 967 (1971).

"Resonance of Coupled Transverse Oscillations in Two Circular Beams", D. G. Koshkarev and P. R. Zenkevich, *Particle Accelerators* **3**, 1 (1972).

"Precession of a Strong Astron E-Layer", R. V. Lovelace and R. N. Sudan, *Phys. Fluids* **15**, 1842 (1972).

"Injection Current Limitation in Astron-like Experiments", A. B. Langdon, M. E. Rensink and T. J. Fessenden, *Plasma Phys.* **15**, 1267 (1973).

"Current Understanding of ERA", L. J. Laslett, *IEEE Trans. Nucl. Sci.* **NS-20**, 271 (1973).

"Envelope Instabilities in Relativistic Electron Rings", D. M. Levine, *IEEE Trans. Nucl. Sci.* **NS-20**, 327 (1973).

"Negative Mass Instability in Hollow Cylinders", D. M. Levine, *IEEE Trans Nucl. Sci.* **NS-20**, 330 (1973).

"Equilibrium and Stability Properties of Relativistic E-Layers and Electron Rings", S. M. Mahajan, Ph. D. thesis, University of Maryland, 1973 [University of Maryland Tech. Rep. No. 74-046].

"Equilibrium Orbit and Linear Oscillations of Charged Particles in Axisymmetric **E X B** Fields and Application to the Electron Ring Accelerator", M. Reiser, *Particle Accelerators* **4**, 239 (1973).

"Beam-Plasma Interaction in Astron", C. D. Striffler and T. Kammash, *Plasma Phys.* **15**, 729 (1973).

"A Computational Study of the Nonlinear Stage of the Development of Radiation Instability in Relativistic Electron Rings", B. G. Shchinov, A. G. Bonch-Osmolovskii, V. G. Makhavkov and V. N. Tsytovich, *Plasma Phys.* **15**, 211 (1973).

"Synchrotron Emission Spectrum for Intense Relativistic Electron Rings", R. C. Davidson and S. M. Mahajan, *Phys. Fluids* **17** in press (1974).

NONNEUTRAL PLASMAS AND SYSTEMS WITH INTENSE SELF FIELDS

The Theory and Design of Electron Beams, J. R. Pierce (Van Nostrand, New York, 1954).

"On the Inertial-Electrostatic Confinement of a Plasma", W. C. Elmore, J. L. Tuck and K. M. Watson, *Phys. Fluids* **2**, 239 (1959).

"Inertial-Electrostatic Confinement of Ionized Fusion Gases", R. L. Hirsch, *J. Appl. Phys.* **38**, 4522 (1967).

"Experimental Studies of a Deep, Negative, Electrostatic Potential Well in Spherical Geometry", R. L. Hirsch, *Phys. Fluids* **11**, 2486 (1968).

"Stability of Crossed-Field Electron Beams", O. Buneman, R. H. Levy, and L. M. Linson, *J. Appl. Phys.* **37**, 3203 (1966).

"Electrostatic Equilibrium of Electron Clouds in Magnetic Fields", R. H. Levy, *Phys. Fluids* **11**, 772 (1968).

"Exact Solution of Poisson's Equation for Space Charge Limited Flow in a Relativistic Planar Diode", J. E. Boers and D. Kelleher, *J. Appl. Phys.* **40**, 2409 (1969).

"Pulsar Electrodynamics", P. Goldreich and W. H. Julian, *Astrophys. J.* **157**, 869 (1969).

"Rigid Drift Model of High-Temperature Plasma Containment", R. L. Morse and J. P. Freidberg, *Phys. Fluids* **13**, 531 (1970).

"Experimental Study of the Magnetic Piston-Shock Wave Problem in a Collisionless Plasma", by A. W. DeSilva, W. F. Dove, I. J. Spalding and G. C. Goldenbaum, *Phys. Fluids* **14**, 42 (1971).

"Investigation of an RF Produced Nonneutral Hot Electron Plasma", N. C. Luhmann, Ph. D. Thesis, University of Maryland, 1972 [University of Maryland Tech. Rep. No. 73–045].

"Stability of Sheared Electron Flow", J. A. Rome and R. J. Briggs, *Phys. Fluids* **15**, 796 (1972).

"Synchrotron Radiation from a Magnetically Confined Nonneutral Hot Electron Plasma", S. F. Nee, A. W. Trivelpiece and R. E. Pechacek, *Phys. Fluids* **16**, 502 (1973).

"Variation of the Magnetic Moment in Nonneutral Electron Plasmas", C. A. Kapetanakos and C. D. Striffler, *Phys. Fluids* **17**, in press (1974).

AUTHOR INDEX

Numbers in parentheses indicate numbers of references cited in the text without authors' names. Numbers set in *italics* designate page numbers on which complete literature citations are given.

Italic page numbers *188-192* refer to "Supplementary References."

SUBJECT INDEX